WILD BERRIES
of ALBERTA, SASKATCHEWAN and MANITOBA

Fiona Hamersley Chambers
Amanda Karst

with contributions from
Andy MacKinnon, Linda Kershaw
John Thor Arnason & Patrick Owen

Lone Pine Publishing

© 2012 by Lone Pine Publishing

First printed in 2012   10 9 8 7 6 5 4 3 2 1

All rights reserved. No part of this work covered by the copyrights hereon may be reproduced or used in any form or by any means—graphic, electronic or mechanical—without the prior written permission of the publisher, except for reviewers, who many quote brief passages. Any request for photocopying, recording, taping or storage on information retrieval systems of any part of this work shall be directed in writing to the publisher.

The Publisher: Lone Pine Publishing
2311 – 96 Street
Edmonton, Alberta   T6N 1G3
Website: www.lonepinepublishing.com

Library and Archives Canada Cataloguing in Publication

Hamersley Chambers, Fiona, 1970–
  Wild berries of Alberta, Saskatchewan and Manitoba / Fiona Hamersley Chambers, Amanda Karst; with contributions from Andy MacKinnon ... [et al.].

Includes bibliographical references and index.
ISBN 978-1-55105-866-5

  1. Berries—Prairie Provinces—Identification. I. Karst, Amanda II. Title.

QK203.P68H36 2012   581.4'6409712   C2011-908462-7

Editorial Director: Nancy Foulds
Project Editor: Gary Whyte
Editorial Support: Kathy van Denderen, Kelsey Attard
Production Manager: Gene Longson
Book Design: Lisa Morley
Layout & Production: Alesha Braitenbach
Cover Design: Gerry Dotto

Photographs and illustrations in this book are used with the generous permission of their copyright holders. Illustration and photo credits are located on p. 183, which constitutes an extension of this copyright page.

Front cover photo: © Victorpr | Dreamstime.com
Back cover photo: © Photos.com

Map Source: Commission for Environmental Cooperation. 2009. Ecological Regions of North America: Level III. CEC. Montreal, Canada.

DISCLAIMER: This guide is not meant to be a "how-to" reference guide for consuming wild berries. We do not recommend experimentation by readers, and we caution that many of the plants in Canada, including some berries, are poisonous and harmful. The authors and publisher are not responsible for the actions of the reader.

We acknowledge the financial support of the Government of Canada through the Canada Book Fund (CBF) for our publishing activities.

PC: 16

Dedications

To three incredible women who I am fortunate to have in my life: my mother Sarah Richardson for being so wonderfully enthusiastic about plants and for doing lots of quality child care while I write, my second mum Vicky Husband for showing me what you can do if you don't take "no" for an answer and persist against the odds, and to Nancy Turner for showing me what you can accomplish by always being positive and kind. And most of all to my boys Hayden and Ben. You guys are the best foragers and berry testers I know and you give me such joy.

–FHC

- To Nancy Turner for her deep respect for ethnobotanical knowledge and special fondness for berry picking;
- To those indigenous people I have been fortunate to work with who have taught me so much about the importance of berries to community;
- To my mom and granny who have taught me about the history of berries in our family;
- And to all the people who gather berries today and keep their uses and traditions alive.

–AK

Acknowledgements

The following people are thanked for their valued contributions to this book:

- The many photographers who allowed us to use their photographs;
- Lone Pine's editorial and production staff;
- And most importantly, the indigenous peoples, settlers, botanists and writers who kept written records or oral accounts of the many uses of the berries featured in this book. We are so grateful to our family, friends and teachers who have shared their knowledge and enthusiasm about our native plants and their uses.

Contents

List of Recipes 5
Plants at a Glance 6
Alberta, Saskatchewan & Manitoba Maps 9

Introduction 12
Edibility 17 • A Cautionary Note 22 • A Few Gathering Tips 24 • Disclaimer 28

THE BERRIES 29

Trees & Shrubs 30
Junipers 30 • Smooth Sumac 36 • Skunkbush 38 • Buckthorns 40 • Elderberries 42 • Bush Cranberries 46 • Red-osier Dogwood 50 • Buffaloberry & Soapberry 52 • Silverberry 54 • Hawthorns 56 • Mountain Ashes 60 • Wild Roses 62 • Plums 66 • Red Cherries 68 • Chokecherry 70 • Saskatoon 72 • Currants 76 • Gooseberries 82 • Raspberries 86 • Thimbleberry 90 • Cloudberry 92 • Bearberries 94 • Black Crowberry 98 • False-wintergreens 100 • Cranberries 104 • Blueberries 108 • Huckleberries 114 • Whortleberry 118 • Riverbank Grape 120

Flowering Plants 122
Sarsaparillas 122 • Common Barberry 124 • Prickly-pear Cacti 126 • Strawberry Blite 128 • Bunchberry 130 • Clintonia & Queen's Cup 132 • Fairybells 134 • Twisted-stalks 136 • Wild Lily-of-the-Valley 140 • False Solomon's-seals 142 • Smooth Solomon's-seal 146 • Strawberries 148 • Northern Comandra 152

Poisonous Plants 154
Western Poison-ivy 154 • Devil's Club 156 • American Bittersweet 158 • Snowberries 160 • Honeysuckles 162 • Canada Moonseed 164 • Red Baneberry 166 • Nightshades 168 • Pacific Yew 170 • False Virginia Creeper 172

Glossary 174
References 177
Index 179
Photo & Illustration Credits 183
About the Authors 184

List of Recipes

Dried Fruit 23
  huckleberries, strawberries, thimbleberries, blueberries, saskatoons or currants

Frozen Wild Fruit the Easy Way 23
  wild berries

Indian Lemonade 37
  sumac flower spikes, frozen blueberries

Indian Ice Cream 53
  soapberries

Rosehip Jelly 65
  rosehips, apples

Pemmican 74
  saskatoons or blueberries

Saskatoon Squares 74
  saskatoons or blueberries

Saskatoon Crumble 74
  saskatoons

Prairie Berry Cordial 89
  juicy berries (raspberries or blueberries)

Wild Berry Dressing 89
  mixed tangy wild berries (raspberries, thimbleberries or huckleberries)

Wild Berry Juice 91
  sweet berries (blueberries, bilberries or thimbleberries)

Berry Fruit Leather 93
  crushed berries (one variety or a mix)

Cranberry Chicken 107
  cranberries

Blueberry Cobbler 113
  blueberries or huckleberries

Fruit Popsicles 113
  wild berries

Huckleberry Relish 116
  huckleberries

Tom's Huckleberry Pie 117
  huckleberries

Riverbank Grape Jelly 121
  riverbank grapes

Wild Berry Muffins 151
  mixed wild fruit (strawberries, thimbleberries, blueberries or huckleberries)

Plants at a Glance

TREES AND SHRUBS

Junipers p. 30

Smooth Sumac p. 36

Skunkbush p. 38

Buckthorns p. 40

Elderberries p. 42

Bush Cranberries p. 46

Red-osier Dogwood p. 50

Buffaloberry and Soapberry p. 52

Silverberry p. 54

Hawthorns p. 56

Mountain Ashes p. 60

Wild Roses p. 62

Plums p. 66

Red Cherries p. 68

Chokecherry p. 70

Saskatoon p. 72

Currants p. 76

Gooseberries p. 82

Raspberries p. 86

Thimbleberry p. 90

PLANTS AT A GLANCE

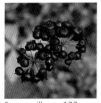
Cloudberry p. 92

Bearberries p. 94

Black Crowberry p. 98

False-wintergreens p. 100

Cranberries p. 104

Blueberries p. 108

Huckleberries p. 114

Whortleberries p. 118

Riverbank Grape p. 120

Flowering Plants

Sarsaparillas p. 122

Common Barberry p. 124

Prickly Pear-cacti p. 126

Strawberry Blite p. 128

Bunchberry p. 130

Clintonia & Queen's Cup p. 132

Fairybells p. 134

PLANTS AT A GLANCE

Twisted-stalks p. 136

Wild Lily-of-the-valley p. 140

False Solomon's-seals p. 142

Smooth Solomon's-seals p. 146

Strawberries p. 148

Northern Comandra p. 152

Poisonous Plants

Western Poison-ivy p. 154

Devil's Club p. 156

American Bittersweet p. 158

Snowberries p. 160

Honeysuckles p. 162

Canada Moonseed p. 164

Red Baneberry p. 166

Nightshades p. 168

Pacific Yew p. 170

False Virginia Creeper p. 172

ALBERTA MAP

Alberta Ecological Regions

- Mid-Boreal Uplands and Peace-Wabaska Lowlands
- Clear Hills and Western Alberta Upland
- Cypress Upland
- Canadian Rockies
- Athabasca Plain and Churchill River Upland
- Hay and Slave River Lowlands
- Coppermine River and Tazin Lake Uplands
- Aspen Parkland/Northern Glaciated Plains
- Northwestern Glaciated Plains

Source: Commission for Environmental Cooperation. 2009. Ecological Regions of North America: Level III. CEC. Montreal, Canada.

SASKATCHEWAN MAP

Saskatchewan Ecological Regions

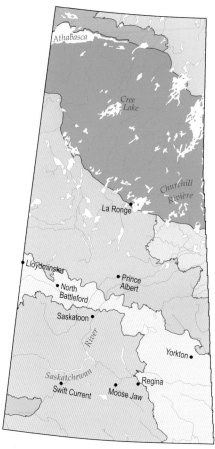

- Mid-Boreal Lowland and Interlake Plain
- Mid-Boreal Uplands and Peace-Wabaska Lowlands
- Cypress Upland
- Athabasca Plain and Churchill River Upland
- Kazan River and Selwyn Lake Uplands
- Coppermine River and Tazin Lake Uplands
- Aspen Parkland/Northern Glaciated Plains
- Northwestern Glaciated Plains

Source: Commission for Environmental Cooperation. 2009. Ecological Regions of North America: Level III. CEC. Montreal, Canada.

Manitoba Ecological Regions

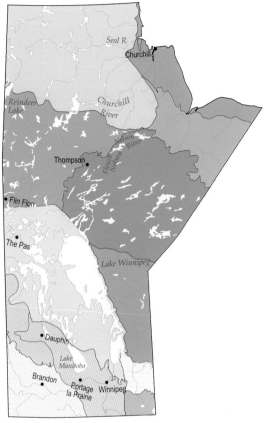

- Mid-Boreal Lowland and Interlake Plain
- Mid-Boreal Uplands and Peace-Wabaska Lowlands
- Northern Lakes and Forests
- Northern Minnesota Wetlands
- Kazan River and Selwyn Lake Uplands
- Aberdeen Plains
- Hudson Bay and James Bay Lowlands
- Coastal Hudson Bay Lowland
- Athabasca Plain and Churchill River Upland
- Lake Nipigon and Lac Seul Upland
- Hayes River Upland and Big Trout Lake
- Aspen Parkland/Northern Glaciated Plains
- Lake Manitoba and Lake Agassiz Plain

Source: Commission for Environmental Cooperation. 2009. Ecological Regions of North America: Level III. CEC. Montreal, Canada.

Introduction

Wood strawberry (*Fragaria vesca*)

It's difficult to find someone who does not enjoy eating a berry. Juicy, sweet, tart, sometimes sour, generally bursting with flavour and very good for you—wild berries are gifts from the land, treasures to be discovered on a casual hike or potentially a lifesaving food if you're unfortunate enough to get lost in the woods. Berries have a long history of human use and enjoyment as food and medicine, in ceremonies and for ornamental and wildlife value. Our ancestors needed to know as a matter of survival which berries were edible or poisonous, where they grew and in which seasons and how to preserve them for non-seasonal use. These early peoples often went to great lengths to manage their wild berry resources: pruning, coppicing, burning, transplanting and even selectively breeding some wild species into the domesticated ancestors of many of our modern fruit varieties.

Today, many of us live in urban environments where the food on our plate and in our pantries comes from great distances away. The first strawberry is no longer an eagerly awaited and delectable harbinger of the summer to come. Rather than a

INTRODUCTION

fleetingly sweet June moment, these fruits are now available on our grocery store shelves almost year-round. A sad result of this convenience and lack of seasonality is that store-bought fruit has little resemblance to its forebears. Grocery store strawberries, for example, are generally not properly ripe, don't taste of much and are not loaded with nutrients. As we become more and more disconnected from our food sources, it is even possible that we are forgetting what a "real" berry tastes like. Perhaps part of the exceptional taste of wild fruit is the thrill of the hunt and the discovery of a gleaming berry treasure hanging—sometimes in great profusion—from a vine or bush. Wild berries are only available for a short time during the year and we must increasingly travel to find them growing in their native state. We must make an effort to discover the berries in the wild or find a reputable source for those plants that will grow in our home gardens.

Wild red raspberry (*Rubus idaeus*)

The wild berries described in this book are, for the most part, not available in stores. When they are, they are often very expensive. A berry in its prime state of ripeness is juicy and delicate, and therefore does not travel well. It is a pity, because slowly savouring these fruits at the peak of perfection plucked fresh off the plant is one of the great joys in life. What better way to spend a warm summer's day than wandering hillsides, country roads, or forest edges with friends and family in search of these delectable morsels? Wild berry gathering builds community and family and is a great way to connect you and your children to nature. In winter, a spoonful of these frozen or preserved wild fruits will bring back the taste of summer for a delicious moment. We hope that, whether you are a seasoned gatherer or a new enthusiast, this book will help guide you to experience and share in this wonderful and generous gift of nature.

Why Learn to Identify and Gather Wild Berries?

Berries gathered in the wild generally have superb flavour and can be gathered when they are properly ripe. These fruits are not only delicious but also contain important nutrients and phytochemicals (anticancer compounds) that are increasingly lacking from our commercially available fruits. Many wild berries are high in vitamin C and also contain trace elements of carbohydrates, proteins and important nutrients such as iron, calcium, thiamine and vitamin A. Although most people will obtain this guidebook to enjoy wild berries on hikes and outings, the information that you learn here could also save your life if you ever get stranded or lost in the backcountry. Be warned, though! Gathering wild berries can also be considered a dangerous "gateway" into the more complex realm of preserving and cooking with these fruits as well as growing them in your own backyard. Once you start on this journey it can become rather addictive and even spread to friends and family!

What Is Not Covered in this Guide

This book should enable you to identify most native berry species in Alberta, Saskatchewan and Manitoba, but it is not intended as a complete reference guide. A section on references and further reading is provided if you wish to study these plants in greater detail. Some berry species are so rarely found or have such a restricted range that it would not be useful to include them here. There are many excellent resources available to help you understand the cultivation and use of domesticated fruit species, so these topics are also not covered. Nuts, seeds and cones are not considered "berries" in the common sense of the word, so they are excluded.

What Is a "Berry"?

In this guide, "berry" is used in the popular sense of the word, rather than in strictly botanical terms, and includes any small fleshy fruit. Technically, a "berry" is a fleshy, simple fruit produced from a single ovary that contains one or more ovule-bearing structures (carpels) that each hold one or more seeds. The outside covering (the endocarp) of a berry is generally soft, moist and fleshy, most often in a globular shape. Roughly translated, a berry is really a seed(s) packaged in a tasty moist pulp that encourages animals to eat the fruit and distribute the seeds far and wide from the parent plant so that these offspring can grow and flourish. "True berries" include currants, huckleberries, blueberries and grapes.

Botanically, however, what we call a "berry" often includes simple fleshy fruits such as drupes and pomes. The botanical definitions of different types of fruit are provided below for general interest and are also sometimes mentioned, where appropriate, in the text.

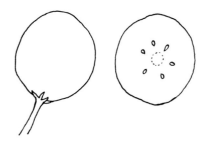

A "drupe" is a fleshy stone fruit that closely resembles a berry but has a single seed or stone with a hard inner ovary wall that is surrounded by a fleshy tissue. Wild fruit in this category include high bush cranberries and bunchberries; some domestic fruit examples are cherries and plums.

INTRODUCTION

Saskatoon berry (*Amelanchier alnifolia*)

A "compound drupe" or "aggregate" fruit ripens from a flower that has multiple pistils, all of which ripen together into a mass of multiple fruits, called "drupelets." A drupelet is a collection of tiny fruit that forms within the same flower from individual ovaries. As a result, these fruit are often crunchy and seedy. Wild examples include raspberries and blackberries. Cultivated examples include tayberries, loganberries and boysenberries.

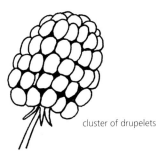

cluster of drupelets

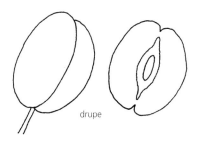

drupe

A "multiple fruit" is similar to an aggregate, but differs in that it ripens from a number of separate flowers that grow closely together, each with its own pistil (as opposed to growing from a single flower with many pistils). Mulberry is the only native Canadian example of a multiple fruit. Tropical examples include pineapples and figs.

An "accessory fruit" is a simple fruit with some of its flesh deriving from a part other than the ripened ovary. In other words, a source other than the

INTRODUCTION

ovary generates the edible part of the fruit. Other names for this type of fruit include pseudocarp, false fruit or spurious fruit. A "pome" is a sort of accessory fruit because it has a fleshy outer layer surrounding a core of seeds enclosed with bony or cartilage-like membranes (the inner core is considered the "true" fruit). Saskatoons (serviceberries) and hawthorns are wild examples of pomes; apples and pears are domestic examples.

Another type of accessory fruit is the strawberry; the main part of the fruit derives from the receptacle (the fleshy part that stays on the plant when you pick a raspberry) rather than from the ovary. Wintergreen is also an example of an accessory fruit type as the fruit is really a dry capsule surrounded by a fleshy calyx.

A "cone" is a fruit made up of scales (sporophylls) that are arranged in a spiral or overlapping pattern around a central core, and in which the seeds develop between the scales. The juniper is an example of a species that produces cones.

A "hip" has a collection of bony seeds (achenes), each of which comes from a single pistil, covered by a fleshy receptacle that is contracted at the mouth. The rose hip (which is also an accessory fruit) is our only example of a hip.

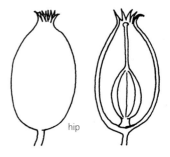

hip

cone

The Species Accounts

In this book, species are organized by growth form into three main sections—Trees and Shrubs, Flowering Plants, and Poisonous Plants. Closely related or similar plants are grouped together for comparison, and the section on poisonous plants is conveniently located at the end of the book.

This book includes plants common in Alberta, Saskatchewan and Manitoba that have been used by people from ancient times to the present. Each account has a detailed description for each plant, including plant form, leaf structure, habitat and range, and fruit form, colour and season. This description, in addition to colour photographs

INTRODUCTION

and illustrations, will help you ensure safe plant and berry identification. Information on traditional and contemporary uses for food, medicines and material culture are also included for general interest.

The information in each species account is presented in an easy-to-follow format. In addition to the opening general discussion, each account includes subheadings of Edibility (see below for edibility scale), Fruit (a description of the look and sometimes the taste of the fruit if it is describable in meaningful terms), Season (flowering and fruiting seasons) and Description (a detailed description of the plant, flowers, habitat and range).

Many species accounts focus on a single species, but if several similar species have been used in the same ways, two or more species may be described together. In these "group" accounts, you will find a general Description for the group, but there will also be separate paragraphs in which the individual species in the group are described in specific detail. So, for example, the Hawthorns (*Crataegus* spp.) account describes historical and modern uses in the opening discussion, then has subheadings of Edibility, Fruit, Season, and Description for all hawthorns. These subheadings are followed by separate paragraphs with important specific information (including identification and location details) for each of the numerous hawthorns in the prairie provinces (black hawthorn, fireberry hawthorn, fleshy hawthorn, red hawthorn).

Where appropriate, you will find an "Also called" subheading that describes other common and scientific names for the species. These "Also called" names are found below the main account title and at the end of individual species description paragraphs within the account.

Edibility

All accounts contain a scale of edibility for each species. Although we have a wonderful variety of native berries, it is useful and interesting to know which ones are worth our time pursuing; which ones, although considered "edible," are better left for the birds or as a famine food; and which ones are toxic or poisonous.

Red hawthorn (*Crataegus columbiana*)

INTRODUCTION

Bunchberry (*Cornus canadensis*)

Highly edible describes those berries that are most delicious and are well worth gathering and consuming. A wild strawberry or saskatoon (serviceberry), for example, is considered highly edible.

Edible describes those berries that are still tasty, but not as good as the prime edible species. An example of such a fruit is bunchberry.

Not palatable describes berries you can eat without any ill effects but are perhaps not worth the effort to harvest given their lack of flavour, their bitterness, relatively large seeds or lack of fleshiness. It is useful to know about these species in case you are desperate to snack on something in the woods, but they are not berries that you would actively gather to make a pie! Silverberry is an example of such a species.

Edible with caution (toxic) are berries that are palatable, but have differing reports as to their edibility, or perhaps they are only toxic if eaten in large amounts, haven't been prepared properly or are unripe. Berries of our native juniper species are an example under this category.

Poisonous berries are ones that are definitely poisonous and should not be eaten. An example of a poisonous berry is red baneberry.

Red baneberry (*Actaea rubra*)

INTRODUCTION

Prairie rose (*Rosa woodsii*)

Season

The season given for flowering and fruit production for each species is an average. Specific microclimates such as deep river valley bottoms or exposed rolling hills will necessarily produce a wide range of flowering and fruiting variability for the same plant. Berry plants produce fruit of differing quality and quantity from year to year, depending on factors such as plant age and health, changes in temperature and moisture, or insect infestation. Some berries, such as rosehips, are best harvested later in the year after the first frost sweetens the fruit.

Description

The plant description and the accompanying photographs and illustrations are important parts of each species account. Each plant description begins with a general outline of the form of the species or genus named at the top of the page. Detailed information about diagnostic features of the leaves, flowers and fruits is then provided. Flowering time is included as part of the flower description to give some idea of when to look for blooms and a general fruiting season is also included. If two or more species of the same genus have been used for similar purposes, several of the most common species may be illustrated and their distinguishing features described.

Plant characteristics such as size, shape, fruiting, colour and hairiness vary with season and habitat and with the genetic variability of each species. Identification can be especially tricky when plants have not yet flowered or fruited. If you are familiar with a species and know its leaves or roots at a glance, you may be able to identify it at any time of year (from very young shoots to the dried remains of last year's plants), but sometimes a positive identification is just not possible.

INTRODUCTION

General habitat information is provided for each species to give you some idea of where to look for a plant. The habitat description provides information about general habitat (e.g., in moist, mossy forest), elevation (e.g., low elevations) and range (e.g., from the northern part of a province to its southern regions). The species ranges and habitats described in this guide were obtained from *Flora of North America*, the United States Department of Agriculture Plants Database website, regional field guides, personal experience and interviews, academic papers and other sources. Despite all due diligence being taken, however, this description is not universal or foolproof. Plants sometimes either grow outside of their reported ranges, or cannot be found within the described habitat. Likewise, the season given for flowering and fruit production for each species is an average. The habitat information included for each species is meant as a general guide only; plants often grow in a variety of habitats over a broad geographical range.

The origin of non-North American species is also noted. The flora of many areas has changed dramatically over the past 200 years, especially in and around human settlements. European settlers brought many plants with them, either accidentally (in ship ballasts, packing and livestock bedding) or purposely (for food, medicine, fibre, ornamental value, etc.). Some of these introduced species produce fruit and have thrived, and some are now considered weeds on disturbed sites across much of Canada. An example of such an introduced species is European mountain ash.

What's In a Name?

Both common and scientific names are included for each plant. Scientific names are from *Flora of North America*, for families already completed (*Flora of North America* is a work in progress). For other plant families, scientific names follow *Flora of Canada* (Scoggan 1978–1979). Common names are also largely from *Flora of North America* and *Flora of Canada*.

Common names are often confusing. Sometimes, the same common name can refer to a number of different, even unrelated, species. And, at the same time, one common name can even refer to a plant that is edible and to a completely different and unrelated species that is poisonous! For this reason, the scientific name is included for each plant entry.

European mountain ash (*Sorbus aucuparia*)

INTRODUCTION

The two-part scientific name used by scientists to identify individual plants may look confusing, but it is a simple and universal system that is worth taking a few moments to learn about. Swedish botanist Carolus Linnaeus, who lived from 1707 to 1778, first suggested a system for grouping organisms into hierarchical categories and it is still essentially the same today, almost 300 years after he first developed it! His system differed from other contemporary ones in that it used an organism's morphology (its form and structure) to categorize a species, with a particular emphasis on the reproductive parts, which we now know are the most ancient part of any plant. Another significant benefit of this hierarchical system is that it groups plants into families so that we can better understand and see how they are related to each other. For example, both oval-leaved blueberry (*Vaccinium ovalifolium*) and lingonberry (*Vaccinium vitis-idaea*) are related cousins in the heath family. At a more distant level, Linnaeus' system shows us that roses are botanically related to apples—both are in the Rosaceae family. Since the names of organisms in Linnaeus' system follow a standard format and are in Latin (or a Latinized name formed from other words), they are the same in every language around the world, making this a truly universal classification and naming protocol.

In Linnaeus' system, the species name (the "scientific name") has two parts: (1) the genus, and; (2) a species identifier (or specific epithet), which is often a descriptive word. The first part of the scientific name, the genus, groups species together that have common characteristics. The genus name is always capitalized and both parts of the scientific name are either written in *italics* or underlined. The second part, the specific epithet, which is not capitalized, often describes the physical or other characteristic of the organism, honours a person, or suggests something about the geographic range of the species. For example, in the scientific name for bunchberry, *Cornus canadensis*, the specific epithet roughly translates as "from Canada." This apt name describes a species that has a wide distribution across our entire country.

It is important to note, however, that botanists do not always agree on how some plants fit into this system. As a result, scientific names can change over time, or there can sometimes be more than one accepted scientific name for a plant. While that is somewhat annoying and may seem redundant, the important thing to remember is that one scientific name

Lingonberry (*Vaccinium vitis-idaea*)

will **never** refer to more than one plant. Thus, if you have identified a wild berry as edible and know its scientific name(s), you can confirm that it is indeed edible and not have to worry that this name may refer to another (possibly deadly poisonous!) plant.

Botanists also sometimes further split species into subsets known as varieties. For example, peaches and nectarines are two slightly different varieties of the peach tree, *Prunus persica*. All cultivated apples are known by the Latin *Malus domestica*. If you purchase a golden delicious apple tree at your local plant nursery, the tag should read "*Malus domestica*, variety Golden Delicious."

Prickly currant (*Ribes lacustre*)

Giving Back to the Plants

While many of our native wild berries grow in profusion, others are threatened by habitat destruction, overharvesting or climate change. In some areas, such as national parks, harvesting is prohibited. Please do not dig up plants from the wild. Most berry species propagate easily from seed or cuttings, and you can also purchase healthy and responsibly produced plants from reputable nurseries. When you harvest native berries in the wild, it's nice to say "thank you" to the plant by weeding back competing species around its base, spreading some of its seeds in similar habitat a short distance from the parent plant, or appropriately pruning the plant if you know the right technique. There is a long history of humans looking after the plants that support us; taking a few moments to continue this tradition and to teach it to our children is time well spent. By learning about our native berry species and harvesting them, we get to know and respect these plants and may even be moved to help protect and propagate them.

A Cautionary Note

If you cannot correctly identify a plant, you should not use it. Identification is more critical with some plants than with others. For example, most people recognize strawberries and raspberries, and all of the species in these two groups are edible, though not all are equally palatable. Plants belonging to the deadly nightshade (*Solanum*) family, however, may be more difficult to distinguish from each other and can range in edibility from highly edible to poisonous. Even the most experienced harvester can sometimes make mistakes. It is important to be certain of a species' identification and any

INTRODUCTION

Recipes

Simple recipes for cooking, preserving or enjoying berries fresh off the plant are included throughout the book. Every berry gatherer or cook can produce delicious results to enjoy with friends and family in the heat of summer and later during the long winter months. The recipes call for specific berries, but you can experiment by substituting other fruit. For example, you could try replacing blueberries with saskatoons, cranberries or huckleberries.

Dried Fruit

It's hard to beat the flavour of home-dried wild berries. Enjoy these special treats out of the bag or add them to your favourite recipes in place of the usual commercial raisins, dried cranberries or blueberries.

A note on berries that dry well: huckleberries, strawberries, thimbleberries (which are a bit crunchy but have a fabulous flavour), blueberries, saskatoons, cranberries and currants. Seedier fruit or very juicy fruit do not preserve as well with this method; it is better to mash these types of berries either alone or combined with other fruit and make them into fruit leather. Some fruit, such as elderberries, should be cooked before drying to neutralize the toxins present in the fresh fruit.

If some of the berries are much larger than others, cut them in half. All the berries on a tray should be roughly the same size to ensure even drying. Carefully pick through the fruit to remove insects and debris. Do not wash the berries—it will cause them to go mushy. Lightly grease a rimmed baking sheet and spread the berries on the sheet so that they do not touch each other. Place in a food dehydrator or dry in an oven at 140° F overnight, leaving the oven door ajar to allow moisture to escape. Cool and store in an airtight container or Ziploc® bag.

Frozen Wild Fruit the Easy Way

Freezing is the quickest and easiest way to preserve wild berries, making a wonderful snack any time of the year. Choose the best and ripest fruit and carefully remove all unwanted debris and insects. Some fruit, such as elderberries, should be cooked first to neutralize any toxins. Give dusty berries a quick rinse, though the extra water and handling may bruise the fruits and stick them together during freezing.

Most instructions tell you to freeze berries individually on rimmed baking sheets before packing them in Ziploc® bags. However, berries can be frozen very successfully in empty milk cartons. Open the carton (1 or 2 L size) fully and wash well in warm, soapy water. Allow to air dry. Cartons with the plastic lid and spout do not work for freezing. Collect cartons during the year, wash them, then store them away for when you need lots of them in the summer. Unless a fruit is particularly mushy (like a very ripe wild raspberry), simply pick through the fruit to clean it, then gently pour the berries into the carton, being careful not to let them pack too hard or crush. Push the top of the carton back together the way it was before opening, then firmly push the top edge so that it folds over flat and indents slightly so that it stays shut. Presto, a sealed container that will never get freezer burn, that is easy to label on the top with a marker pen, and that stacks beautifully in the deep freeze!

To get the frozen fruit out, gently squeeze the carton sides to separate the berries, making it easy to pour out the desired quantity before resealing the carton and returning it to the freezer. If the berries are more firmly attached, simply place the carton on the floor and gently stand on it, turning the sides a few times. As a last resort, peel the carton down to the desired level, cut off the exposed fruit chunk with a sharp knife, and put the remainder of the carton in a Ziploc® bag before replacing it in the freezer. You can even re-use the same cartons for many years as long as the fruit inside is not too mushy or difficult to extract.

INTRODUCTION

Northern gooseberry (*Ribes oxyacanthoides*)

A Few Gathering Tips

1. Gather only species that are common and abundant, and never take all the fruit off one plant. Even then, a cautious personal quota will still deplete the plants if too many people gather them in one area. Remember, plants growing in harsh environments (e.g., northern areas, alpine, desert) might not have enough energy to produce flowers and fruits every year. Also, don't forget the local wildlife. Survival of many animals can depend on access to the fruits that you are harvesting.

2. Never gather from plants that grow in protected and/or heavily used areas such as parks and nature preserves. Doing so is not only wrong, but is also often illegal. Be sure to check the regulations for the area you are visiting.

3. Take only what you need, and damage the plant as little as possible. If you want to grow a plant in your garden, try propagating it from seed or a small cutting rather than transplanting it from the wild.

4. Don't take more than you will use. If you are gathering a plant for food, taste a sample. You may not like the taste of the berries, or the fruit at this site may not be as sweet and juicy as the ones you gathered last year.

5. Gather berries only when you are certain of their identity. Many irritating and poisonous plants grow wild in Canada, and some of them resemble edible or medicinal species. If you are not positive that you have the right plant, don't use it. It is better to eat nothing at all than to be poisoned!

INTRODUCTION

special treatment required before eating a wild berry. Chokecherries, for example, are best cooked to neutralize the poisonous cyanide compounds found in their seeds, and many types of under-ripe berries can cause digestive upset or even be poisonous. Some rare individuals have an allergic reaction to certain berry species.

It is also important to know which parts of the berry are edible. For example, while the fleshy "berry" (it is really an "aril") of the Pacific yew tree is considered edible, eating this fruit is not recommended as the small hard seed contained inside is so deadly poisonous that ingesting even a few can cause death! As a general rule, most of our native berry species taste good and are edible. Those berries that have a bitter, astringent or unpalatable taste are telling us that they are toxic or poisonous and that we should not be eating them. These species tend to rely on birds, rather than humans, to eat the fruit and distribute the seeds. The exceptions to these guidelines are the many introduced ornamental plants in our gardens and municipal plantings, some of which have naturalized into the wild and have sweet-tasting fruit. It is not recommended that you sample these non-native fruits without a positive identification. An example of a common poisonous berry is the European lily-of-the-valley.

Finally, some people believe that it is OK to eat berries that they see birds and wildlife enjoying. That is simply not the case! Do not test this flawed bit of folklore! Likewise, the fact that a plant has edible fruit does not mean that the plant itself is edible.

Pay attention to where you are harvesting. Fruit growing along the edge of a busy highway or near an industrial area could be contaminated with heavy metals or other pollutants. Municipal plantings might look delicious, but they may be sprayed with pesticides and you might not be welcome to harvest the fruit if it has an ornamental value. Please also remember to harvest on public, not private, lands unless you have received permission from the property owner.

Many plants have developed very effective protective mechanisms. Thorns and stinging hairs discourage animals from touching, let alone eating, many plants. Bitter, often irritating and poisonous compounds in leaves and roots repel grazing animals. Many protective devices are dangerous to humans. The "Warning" boxes throughout this book include notes of potential hazards associated with the plant(s) described. Hazards can range from deadly poisons to

Pacific yew (*Taxus brevifolia*)

INTRODUCTION

Common juniper (*Juniperus communis*)

spines with irritating compounds in them. These "Warning" boxes may also describe poisonous plants that could be confused with the species being discussed in the account.

The fine line between delicious and dangerous is not always clearly defined. Many of the plants that we eat every day contain toxins, and almost any food is toxic if you eat too much of it. Personal sensitivities can also be important. People with allergies may die from eating common foods (e.g., peanuts) that are harmless to most of the population. Most wild plants are not widely used today, so their effects on a broad spectrum of society remain unknown.

As with many aspects of life, the best approach is "moderation in all things." Sample wisely—when trying something for the first time, take only a small amount to see how you like it and how your body reacts.

No Two Plants Are the Same

Wild plants are highly variable. No two individuals of a species are identical and many characteristics can vary. Some of the more easily observed characteristics include the colour, shape and size of stems, leaves, flowers and fruits. Other less obvious features, such as sweetness, toughness, juiciness and concentrations of toxins or drugs, also vary from one plant to the next.

Many factors control plant characteristics. Time is one of the most obvious. All plants change as they grow and age. Usually, young leaves are the most tender, and mature fruits are the largest and sweetest. Underground structures also change throughout the year.

Habitat also has a strong influence on plant growth. The leaves of plants from moist, shady sites are often larger, sweeter and more tender than those of plants on dry, sunny hillsides. Berries may be plump and juicy one year, when shrubs have had plenty of moisture, but they can become dry and wizened during a drought. Without the proper nutrients and environmental conditions, plants cannot grow and mature.

Finally, the genetic make-up of a species determines how the plant develops and how it responds to its environment. Wild plant populations tend to be much more variable than domestic crops, partly because of their wide range of habitats, but also because of their greater genetic variability. Humans have been planting and harvesting plants for millennia, repeatedly selecting and breeding plants with the most desirable characteristics. This process has produced many highly productive cultivars—trees with larger, sweeter fruits, potatoes with bigger tubers and sunflowers with larger, oilier seeds. These crop species are more productive, and they also produce a specific product each time they are planted. Wild plants are much less reliable.

Wild species have developed from a broader range of ancestors growing in many different environments, so their genetic make-up is much more variable than that of domestic cultivars. One population may produce sweet, juicy berries while the berries of another population are small and tart; one plant may have low concentrations of a toxin that is plentiful in its neighbour. This variability makes wild plants much more resilient to change. Although their lack of stability may seem to reduce their value as crop species, it is one of their most valuable features. Domestic crops often have few defences and must be protected from competition and predation. As fungi, weeds and insects continue to develop immunities to pesticides, we repeatedly return to wild plants for

Pin cherry (*Prunus pensylvanica*)

Chokecherry (*Prunus virginiana*)

INTRODUCTION

Red-osier dogwood (*Cornus sericea*)

new repellents and, more recently, for pest-resistant genes for our crop plants.

Disclaimer

This book summarizes interesting, publicly available information about many plants in Canada. It is not intended as a "how-to" guide for living off the land. Rather, it is a guide for people wanting to discover the astonishing biodiversity of our useful plants and to connect to our cultural traditions, especially those of the First Nations. Only some of the most widely used species in Canada, and only some of their uses, are described and discussed. We do not recommend self-medication with herbal medicines. Using plant medicines and consumption of wild foods should only be considered under guidance from an experienced healer/elder/herbalist. As a field guide, the information presented here is limited, and further study of species of interest should be made using other botanical literature. No plant or plant extract should be consumed unless you are absolutely certain of its identity and toxicity and of your personal potential for allergic reactions. The authors and publisher are not responsible for the actions of the reader.

Twinflower honeysuckle (*Lonicera involucrata*)

THE BERRIES

Junipers *Juniperus* spp.

Rocky mountain juniper (*J. scopulorum*)

Some tribes cooked juniper berries into a mush and dried them in cakes for winter use. The berries were also dried whole and ground into a meal that was used to make cakes. In times of famine, small pieces of the bitter bark or a few berries were chewed to suppress hunger. Dried, roasted juniper berries have been ground and used as a coffee substitute, and teas were occasionally made from the stems, leaves and/or berries, but they were usually used as medicine rather than a beverage.

Juniper berries are well known for their use as a flavouring for gin, beer and other alcoholic drinks. Tricky Marys can be made by soaking juniper berries in tomato juice for a few days and then following your usual recipe for Bloody Marys, but omitting the alcohol. The taste is identical, and the drink is non-alcoholic.

Juniper berries can be quite sweet by the end of their second summer or in the following spring, but they have a rather strong, "pitchy" flavour that some people find distasteful. They can be added as flavouring in meat dishes (recommended for venison and other wild game, veal and lamb), soups and stews, either whole, crushed or ground and used like pepper. Rocky mountain juniper sprigs were also sometimes placed among dried salmon or other stored foods to protect them against attack from insects and flies.

Juniper berry tea has been used to aid digestion, stimulate appetite, relieve colic and water retention, treat diarrhea, vomiting and heart, lung and kidney problems, stop bleeding, reduce swelling and inflammation and calm hyperactivity. The berries were also chewed as a "cure-all medicine" (only one berry eaten), to relieve cold symptoms, to settle upset stomachs and to increase appetite. They have been smoked in a pipe to treat asthma. Oil-of-juniper (made from the berries) was mixed with fat to make a salve to protect wounds from irritation by flies. Juniper berries have been used as a diuretic; they are reported to stimulate urination by irritating the kidneys and will give the urine a violet-like fragrance. They are also said to stimulate sweating, mucous secretion, production of hydrochloric acid in the stomach and contractions in the uterus and intestines.

Some studies have shown juniper berries to lower blood sugar caused by adrenaline hyperglycemia, suggesting that they may be useful in the

Common juniper (*J. communis*)

Common juniper (*J. communis*)

Junipers

treatment of insulin-dependent diabetes. Juniper berries also have antiseptic qualities, and studies by the National Cancer Institute have shown that some junipers contain antibiotic compounds that are active against tumours. Strong juniper tea has been used to sterilize needles and bandages, and during the Black Death in 14th-century Europe, doctors held a few berries in the mouth when treating patients as they believed that this would prevent them from being infected. During cholera epidemics in North America, some people drank and bathed in juniper tea to avoid infection. Juniper tea has been given to women in labour to speed delivery, and after the birth, it has been used as a cleansing, healing agent.

Juniper berries were sometimes dried on strings, smoked over a greasy fire and polished to make shiny black beads for necklaces or headpieces.

Common juniper (*J. communis*)

Some tribes also scattered berries on anthills—the ants would eat out the sweet centre, leaving a convenient hole for stringing the beads to make necklaces. Smoke from the berries or branches of junipers has been used in religious ceremonies or to bring good luck (especially for hunters) or protection from disease, evil spirits, witches, thunder, lightning and so on.

The berries make a pleasant, aromatic addition to potpourris, and vapours from boiling juniper berries in water were used to purify and deodorize homes affected by sickness or death. They can also be used to make a brown dye. These plants are decorative, particularly in the winter months, and make a hardy and drought-tolerant addition to the ornamental garden. Junipers can be very long-lived, with some recorded specimens as old as 1500 years!

EDIBILITY: edible, but with caution

FRUIT: Small fleshy cones ("berries") are ripe when bluish-purple to bluish-green colour.

SEASON: Berries form on female plants only and mature the following year but are present on the plant all year round.

DESCRIPTION: Coniferous, evergreen shrubs or small trees, to 20 m tall, with some species creeping low on the ground. Leaves scale-like, 1.5–15 mm long, in rows, dark green to yellowy. Male plants produce yellow pollen on cones 5 mm long. Females bear small berries, 5–10 mm wide, first green then maturing to a bluish-purple colour. Grows on open, dry rocky areas and grasslands.

TREES & SHRUBS

Common juniper (*J. communis*)

Creeping juniper (*J. horizontalis*)

Common juniper (*J. communis*)

Common juniper (*J. communis*) grows to 1 m tall but is normally lower than this. Growth habit is branching, prostrate, trailing, forming wide mats 1–3 m in size. Leaves are needle-like to narrowly lance-shaped, dark green below, whitish above, prickly, 5–15 mm long, in whorls of 3. Bark reddish brown, scaly, thin, shredding. Grows on dry, open sites and forest edges, gravelly ridges and

Junipers

muskeg from lowland bogs to plains and subalpine zones throughout the prairie provinces. Also called: ground juniper.

Creeping juniper (*J. horizontalis*) is a low shrub (seldom over 25 cm tall) with trailing branches. Leaves scale-like, tiny (1.5 mm), in 4 vertical rows, lying flat against the branch, green to grey-green or bluish. Grows in dry, rocky soils in sterile pastures and fields throughout the prairie provinces.

Rocky Mountain juniper (*J. scopulorum*) is a rarer species and grows to 15 m tall. Leaves opposite, 5–7 mm

Creeping juniper (*J. horizontalis*)

long, in 4 vertical rows, young leaves often needle-like, but mature leaves tiny and scale-like. Grows on dry, rocky ridges, open foothills, grasslands and bluffs from AB to southwestern SK.

Rocky Mountain juniper (*J. scopulorum*)

TREES & SHRUBS

Rocky Mountain juniper (*J. scopulorum*)

WARNING: *Some reports consider juniper berries to be poisonous and the leaves and oil of all junipers are toxic. While the addition of a few berries to flavour a dish is likely safe, eating these berries regularly or in large quantities is not advised. Large and/or frequent doses of juniper can result in convulsions, kidney failure and an irritated digestive tract. People with kidney problems and pregnant women should never take any part of a juniper internally. Juniper oil is strong and can cause blistering.*

Rocky Mountain juniper (*J. scopulorum*)

Smooth Sumac *Rhus glabra*

Smooth sumac (*R. glabra*)

Sumac's showy red fruit clusters taste lemony-sour and are beautiful to look at. The fruits have been used to make drinks, jellies or lemon pies. For example, fresh berries can be steeped in water and sweetened to make a cool beverage like lemonade, and dried berries can be cooked in water with maple sugar for a hot beverage. When chewed as a trail nibble, sumac fruits relieve thirst and leave a pleasant taste in the mouth. The tangy lemon flavour of sumac fruit (which really comes from the hairs covering the seeds) is a common ingredient in some Middle Eastern dishes.

The fruits were boiled to make a wash to stop bleeding after childbirth. The berries, steeped in hot water, made a medicinal tea for treating diabetes, bowel problems and fevers. This tea was also used as a wash for ringworm, ulcers and skin diseases such as eczema. The bark and berries have also been used in Ojibway medicine ceremonies.

Sumac is a very decorative and hardy species that provides an interesting fall and winter garden display. It does tend to sucker, though, so can get invasive in the garden if not kept in check.

EDIBILITY: edible

FRUIT: Reddish orange, densely hairy, sticky, berry-like drupes, 4–5 mm long, in persistent, fuzzy clusters.

SEASON: Flowers May to July. Fruits ripen July to August, often remaining through winter.

DESCRIPTION: Deciduous shrub or small tree, 1–3 m tall, usually forming thickets. Twigs and leaves are hairless; buds with whitish hairs. Branches exude milky juice when broken. Leaves pinnately divided into 11–31 lance-shaped, 5–12 cm long, toothed leaflets, turning bright red in autumn. Flowers cream-coloured to greenish yellow, about 3 mm across, with 5 fuzzy petals, forming dense, pyramid-shaped, 10–25 cm long clusters. Grows across the prairie provinces on dry forest openings, prairies, fencerows, roadsides and burned areas.

Indian Lemonade

Makes 8½ cups

Enhance this beautiful pink "lemonade" by adding ice cubes (or frozen blueberries) and green mint sprigs.

3 cups dried and crumbled sumac flower spikes
8 cups water • sugar to taste

Pick through the dried flower spikes to remove any dirt or debris. Crumble the red "berries" off the main spike and place them in a jug. Pour the cold water over the berries, mash the mixture with a wooden spoon or potato masher, then let sit for at least an hour. *Do not heat this mixture because it will alter the taste of the sumac.* Strain the liquid through a cheesecloth, jellybag or fine-mesh sieve, and add sugar to taste.

Skunkbush *Rhus trilobata*

Skunkbush (*R. trilobata*)

Skunkbush fruits were eaten raw (sometimes ground with a little water) or boiled, or they were dried and ground into meal. The berries and meal were often mixed with other foods, especially sugar and roasted corn. Skunkbush berries can be crushed, soaked in cold water and strained to make a pink lemonade-like drink. Soaking in hot water, however, produces a bitter-tasting liquid as the heat releases tannins. The inner bark was also eaten as a food.

The leaves were chewed to cure stomachaches or boiled to make a contraceptive tea that was said to induce impotence. Tea made by boiling the root bark was taken to aid delivery of the afterbirth, and skunkbush leaves were used in poultices to relieve itching. The fruits were boiled and their oil, skimmed from the surface, was used to prevent hair loss. The dried berries were ground and applied to smallpox wounds. The leaves were boiled to produce a

TREES & SHRUBS

black-coloured dye for baskets, leather and wool, and the red fruits were boiled with another plant to make red-brown dye. Both ashes and the fruits were used as a mordant to set dyes.

EDIBILITY: edible

FRUIT: Sticky, fuzzy, reddish orange, berry-like drupes, 6–8 mm long, in small clusters.

SEASON: Flowers May to June. Fruits ripen June to October.

DESCRIPTION: Strong-smelling, 1–2 m tall shrub with dark grey-green leaves, alternate, divided into 3 broad-tipped, lobed leaflets that taper to wedge-shaped bases. Flowers yellowish green, under 2 mm across, with 5 fuzzy petals, forming close clusters of spikes near the branch tips, before the leaves appear. Grows along streams and in open areas in prairies, shrublands and foothills in southern AB and SK.

Buckthorns   *Rhamnus* spp.

Alder-leaved buckthorn (*R. alnifolia*)

Some sources report that the purple berries of these small trees or shrubs were eaten by Native peoples, but possibly in modest amounts given their strong purgative effects. The bark, leaves and fruits of alder-leaved and European buckthorn cause vomiting and diarrhea when ingested (see Warning).

These species have been used as sources of laxatives for hundreds of years. Indigenous peoples collected the bark in spring and summer and then dried and stored it for later use. Ingesting fresh bark and berries can have very severe effects, but curing the bark for at least 1 year or using a heat treatment reduces the harshness. Some tribes used these plants as a purgative to induce vomiting when poisons had been eaten. A poultice from the plant was also applied to swelling caused by poison. The chewed bark was used for children with worms, to treat rheumatism, arthritis, and itch, and as a blood purifier, tonic, and physic. The plant was also used as a wash to treat sore or inflamed eyes. European buckthorn is an introduced species that has escaped cultivation and become naturalized in some areas.

European buckthorn (*R. cathartica*)

EDIBILITY: edible with caution (toxic)

FRUIT: Berry-like drupes 6–10 mm long, often ripening unevenly in bunches, turning from green to yellow to a purple or bluish-black colour.

SEASON: Flowers June to July. Fruits ripen August to September.

DESCRIPTION: Erect or spreading shrubs or small trees, 0.5–6 m tall, with alternate, oval to elliptic, prominently veined leaves. Flowers inconspicuous, greenish yellow, all male or all female in 1 cluster (usually sexes on separate plants), forming flat-topped clusters in lower leaf axils.

Alder-leaved buckthorn (*R. alnifolia*) is a medium-sized shrub, 0.5–1.5 m tall, with 2–5 stalkless flower clusters. Leaves 2–10 cm long, 1–5 cm wide, oval, short-stalked, margins toothed or wavy, 6–8 veined. Grows in moist, open to shady meadows, swamps and on streambanks in all prairie provinces.

European buckthorn (*R. cathartica*) is generally a small tree, to 6 m tall. Branches grey and tipped with a small thorn. Leaves 3–6 cm long, 1–4 cm wide, oval to elliptic lateral veins in 3 pairs, margins smooth. Inhabits streambanks, roadsides and waste areas where it has escaped from cultivation and naturalized.

> **WARNING:** *This genus contains large amounts of anthraquinones, which are responsible for its emetic properties. The berries cause vomiting and diarrhea, so are not recommended for consumption. The plants are also best avoided when gathering sticks for roasting wieners.*

Alder-leaved buckthorn (*R. alnifolia*)

European buckthorn (*R. cathartica*)

Elderberries   *Sambucus* spp.

Blue elderberry (*S. nigra* ssp. *cerulea*)

Raw elderberries are generally considered inedible and cooked berries edible (see Warning), but some tribes are said to have eaten large quantities fresh from the bush. Cooking or drying destroys the rank-smelling, toxic compounds. Most elderberries were consumed after steaming or boiling, or were dried for winter use. Sometimes clusters of fruit were spread on beds of pine needles in late autumn and covered with more needles and eventually with an insulating layer of snow. These caches were easily located in the winter months by the bluish-pink stain they left in the snow! Only small amounts of the fruit were eaten at a time, though, just enough to get a taste. Today, they are used in jams, jellies, syrups, preserves, pies and wine. Because these fruits contain no pectin, they are often mixed with tart, pectin-rich fruits such as crab apples. Elderberries are also used to make teas and to flavour some wines (e.g., Liebfraumilch) and liqueurs (e.g., Sambuca). A delicious, refreshing fizzy drink called elderflower pressé or cordial can be made from the flowers. The flowers can also be used to make tea or wine, and in some areas, flower clusters were popular dipped in batter and fried as fritters or stripped from their relatively bitter stalks and mixed into pancake batter.

Red elderberry (*S. racemosa* var. *pubens*)

Blue elderberry (*S. nigra* ssp. *cerulea*)

Red elderberry (*S. racemosa* var. *pubens*)

Elderberries are rich in vitamin A, vitamin C, calcium, potassium and iron. They have also been shown to contain antiviral compounds that could be useful in treating influenza. The berries can be used to produce a brilliant crimson- or violet-coloured dye. Elderberry wine, elderberries soaked in buttermilk and elderflower water have all been used in cosmetic washes and skin creams. The scientific name for *Sambucus* derives from the Greek instrument *sambuke*, in reference to the hollow pithy stems of

Elderberries

Red elderberry (*S. racemosa* var. *pubens*)

Blue elderberry (*S. nigra* ssp. *cerulea*)

Red elderberry (*S. racemosa* var. *pubens*)

this plant, which have been used in many different cultures to make musical instruments.

EDIBILITY: edible, edible with caution (toxic)

FRUIT: Fruits juicy, berry-like drupes, 4–6 mm across, in dense, showy clusters.

SEASON: Flowers April to July. Fruits ripen July to September.

DESCRIPTION: Unpleasant-smelling, 1–3 m tall, deciduous shrubs with pithy, opposite branches often sprouting from the base. Leaves pinnately divided into 5–9 sharply toothed leaflets about 5–15 cm long. Flowers white, 2–6 mm wide, forming crowded, branched clusters.

Blue elderberry (*S. nigra* ssp. c*erulea*) has flat-topped flower clusters and dull blue fruits with a whitish bloom. Usually 9 leaflets. Grows in gravelly, dry soils on streambanks, field edges and woodlands in AB. Also called: *S. glauca, S. cerulea.*

Red elderberry (*S. racemosa*) has pyramid-shaped flower clusters and shiny fruits, and is considered the

TREES & SHRUBS

Red elderberry (*S. racemosa* var. *pubens*)

Blue elderberry (*S. nigra* ssp. *cerulea*)

tastiest of the genus. There are two common subspecies of *S. racemosa*: black elderberry or Rocky Mountain elderberry (*S. racemosa* var. *melanocarpa*) has purplish black fruit; and red elderberry (*S. racemosa* var. *pubens*) bears red fruit. Both grow in open woods, forest edges, moist meadows and roadsides on montane and subalpine sites across the prairie provinces.

> **WARNING:** *All parts of elderberry plants, except for the fruit and flowers, are considered toxic. The stems, bark, leaves and roots contain poisonous cyanide-producing glycosides (especially when fresh) that cause nausea, vomiting and diarrhea, but the ripe fruits and flowers are edible. The seeds, however, contain toxins that are most concentrated in red-fruited species. Many sources classify red-fruited elderberries as poisonous and black- or blue-fruited species as edible.*

45

Bush Cranberries *Viburnum* spp.

High bush cranberry (*V. edule*)

Raw bush cranberries are high in vitamin C and can be very sour and acidic (much like true cranberries), but many Native peoples ate them. One preferred method was to chew the fruit, swallow the juice and then spit out the tough skins and seeds. They were also eaten with bear grease, or in an early year they could be mixed with sweeter berries such as saskatoons. The berries were sometimes added to pemmican if other sweeter berries (e.g., blueberries, saskatoons) weren't available. Berries were traditionally mixed with grease or water and stored in birchbark baskets in an underground cache for winter use, for trade or as a valuable gift.

Bush cranberries are an excellent winter-survival food because they remain on the branches all winter and are high enough that they don't get covered by snow. Berries are best picked in autumn, after they have been softened and sweetened by a frost. Traditionally, they were sometimes picked in mid-winter or spring to take advantage of the improved flavour. Today, bush cranberries are usually boiled, strained (to remove the seeds and skins) and used in jams and jellies.

WARNING: *Some sources classify raw bush cranberries as poisonous, while others report that they were commonly eaten by native peoples. A few berries may be harmless, but ingesting large quantities can cause vomiting and cramps, especially if they are not fully ripe, so it is probably best to cook the fruit before eating. Despite the common name "cranberry," these species are not botanically related to the sour red berries we traditionally enjoy with a Thanksgiving feast.*

TREES & SHRUBS

High bush cranberry (*V. edule*)

EDIBILITY: edible, edible with caution (toxic)

FRUIT: Juicy, strong-smelling, red to orange berry-like drupes 8–10 mm long, with a single flat stone.

SEASON: Flowers April to July. Fruits ripen September to October.

DESCRIPTION: Deciduous shrubs with opposite, 3-lobed leaves, 3–12 cm long. Flowers white, small, 5-petalled, forming flat-topped clusters rather like a lace-cap hydrangea. Leaves turn a showy red colour in fall.

High bush cranberry (*V. edule*) is a scraggly looking shrub 0.5–2 m tall. Bark smooth, grey with a reddish tinge. Leaves opposite, 4–10 cm wide, sharply toothed and hairy underneath, often with 3 shallow lobes evident toward the leaf tip. Flowers small, in clusters 1–3 cm wide, growing beneath leaf pairs. The plant's distinctive, musty smell may announce its presence before it is actually seen. Inhabits shady foothills, damp woods, streambank thickets, and some montane and

While these preserves usually require additional pectin (especially after the berries have been frozen), there are reports that imperfectly ripe berries (not yet red) gel without added pectin. Some people compare their fragrance to that of dirty socks, but the flavour is good (perhaps a Stilton of the berry world?). The addition of lemon or orange peel to the cooking fruit, however, is said to eliminate this odour.

Bush cranberry juice can be used to make a refreshing cold drink or fermented to make wine, and the fresh or dried berries can be steeped in hot water to make tea. Unfortunately, their large stones and tough skins limit their use in muffins, pancakes and pies without the fruit being cooked and strained first. Ripe fruits have been boiled to make a cough medicine. The berries produce a lovely reddish-pink dye, and the acidic juice can be used as a mordant (required to set dyes and make the colour permanent).

American bush cranberry makes a wonderful garden ornamental that is drought tolerant and provides not only pretty and scented spring flowers but also a showy fall foliar display and important winter wildlife food (if the humans don't get there first!).

High bush cranberry (*V. edule*)

47

Bush Cranberries

Nannyberry (*V. lentago*)

Nannyberry (*V. lentago*)

subalpine sites in all prairie provinces. Also called: mooseberry, squashberry, low bush cranberry.

Nannyberry (*V. lentago*) is a shrub, to 6 m tall, with unlobed, pinnately veined leaves, long, tapering leaf tips and winged petioles. The plant has an unpleasant, goat-like smell. Fruit blue-black, raisin-like. Grows in rich soils along woodland edges, streams and rocky hillsides in SK and MB.

American bush cranberry (*V. opulus*) is a large 1–4 m tall shrub with a wide-spreading habit. Leaves maple leaf–like with 3 relatively deeply cut lobes. Flower clusters 5–15 cm across, with a showy outer ring of large (1–2 cm wide), white, sterile flowers surrounding a central growth of tiny (3–4 mm across) petal-less blooms. Grows in moist soils, hedges, scrub areas, plains and woodlands in all prairie provinces. Also called: high bush cranberry (note that this is the same common name sometimes attributed to *V. edule*) • *V. trilobum* var. *americanum*.

American bush cranberry (*V. opulus*)

American bush cranberry (*V. opulus*)

TREES & SHRUBS

Nannyberry (*V. lentago*)

American bush cranberry (*V. opulus*)

High bush cranberry (*V. edule*)

American bush cranberry (*V. opulus*)

49

Red-osier Dogwood *Cornus sericea*
Also called: western dogwood, red willow • *C. alba, C. stolonifera, Svida sericea*

Red-osier dogwood (*C. sericea*)

The fruits of red-osier dogwood are definitely tart and bitter for modern-day tastes. The berries of red-osier dogwood, despite their bitterness, were gathered by some First Nations people in late summer and autumn and eaten immediately. They were also occasionally stored for winter use, either alone or mashed with sweeter fruits such as saskatoons, and in more modern times with sugar. The red-osier dogwood fruit can be cooked when fresh to release the juice, which purportedly makes a refreshing drink when sweetened. Some people separated the stones from the mashed flesh and saved them for later use. They were then eaten as a snack, somewhat like peanuts are today, but this is not recommended in large quantities, and the taste is probably not worth the effort involved.

The fruit has been used to make a wash to treat snow blindness and as a treatment for tuberculosis. Hunters might eat a few berries for luck before they go on a bear hunt (bears are said to be fond of these berries).

Red-osier dogwood is a popular and attractive ornamental tree with good wildlife and aesthetic values. It is also a great choice for making decorative and functional twig furniture, especially when branches range from red to dark green in colour.

EDIBILITY: edible

FRUIT: White (sometimes bluish), pea-sized berry-like drupe, 5–6 mm across, containing a large, flattened stone. Although bitter tasting, it is reported that the whiter fruits are less bitter than the bluer-tinged ones.

SEASON: Flowers May to June. Berries ripen late summer, are edible when white or bluish coloured, and traditionally gathered from August to October.

DESCRIPTION: Erect to sprawling, slender deciduous shrub or small tree, slender and branching in form, usually 2–3 m tall. Twigs and branches are opposite, shiny and smooth, conspicuously bright green to red (the more sun exposure on the stem, the brighter the red colour) when young, changing to brown when older. Leaves opposite, simple, pointed, toothless, 2–8 cm long, with 5–7 prominent parallel leaf veins following the smooth leaf edges toward the tips, greenish above, white to greyish below, becoming red in autumn. Flowers small, white, in flat-topped clusters, 2–5 cm wide, at branch tips. Grows on moist sites, shores and thickets throughout the prairie provinces.

WARNING: *All parts of this species are considered toxic if consumed in large quantities.*

Buffaloberry & Soapberry *Shepherdia* spp.

Buffaloberry (*Shepherdia* spp.)

Soapberries have a bitter flavour that is an acquired taste; however, they were an important fruit for some First Nations. The Blackfoot and other Plains groups ate buffaloberries fresh or boiled, or they were formed into cakes and dried over a fire for winter use. The berries were also cooked, sometimes mixed with sugar, fried in animal fat, mixed with moose liver, added to stews, made into syrup or used to flavour bison meat. The fruit is reportedly more palatable after a frost. Today they are used to make jams and jellies. They also were made into a juice; canned soapberry juice, mixed with sugar and water, makes a refreshing "lemonade." Although they are bitter, soapberries can be used in moderation as an emergency food, since they stay on the shrubs for a long time and are easy to find (see Warning).

Because their juice is rich in saponin, soapberries become foamy when beaten. The ripe fruit was mixed about 4:1 with water and whipped like egg whites to make a foamy dessert called "Indian ice cream" by indigenous groups in Alberta and BC. The resulting foam is truly unexpected and remarkable, having a beautiful white to pale pink colour and the smooth shiny consistency of the best whipped meringue! The incredibly thick foam is rather bitter, so it was usually sweetened with sugar or with other berries. Traditionally, this dessert was beaten by hand or with a special stick with grass or strands of bark tied to one end. These tools were eventually replaced by egg-beaters and mixers. Like egg whites, soapberries will not foam in plastic or greasy containers. The fruit was collected by beating the branches over a canvas or hide and then rolling the berries down a wooden board into a container to separate leaves and other debris.

Soapberries can be crushed or boiled to use as a liquid soap. The red berries were also used to make a red dye.

WARNING: *Both species contain saponin, a bitter, soapy substance that can irritate the stomach and cause diarrhea, vomiting and cramps if consumed in large amounts.*

Soapberries are rich in vitamin C and iron. They have been taken to treat flu, stomachaches, heart problems, arthritis, tuberculosis and indigestion and have been made into a medicinal tea for relieving constipation. Canned soapberry juice, mixed with sugar and water, has been used to treat acne, boils, digestive problems and gallstones.

EDIBILITY: edible

FRUIT: Juicy, bright red oval berries with a fine silvery scale.

SEASON: Flowers April to May. Fruits ripen July to September.

DESCRIPTION: Deciduous shrubs with opposite leaves, smooth edged, 2.5–5 cm long. Flowers yellowish to greenish, about 2 mm, male and female flowers on separate plants, single or forming small clusters at leaf axils.

Silver buffaloberry (*S. argentea*) is a tall shrub, 2–4 m tall, with rough spines along silver branches, young twigs covered in a silvery "scurf." Leaves oblong to ovate, silvery scurfy on the undersides. Grows in open woods, thickets, and hillsides from southern AB to southern MB.

Soapberry (*S. canadensis*) grows to 3 m tall. Young twigs covered in a brown or rusty scale. Older twigs and branches brownish-red with orange flecks, sometimes fissured. Leaves somewhat thick, elliptic, tip rounded, top green with short silvery scales, rusty underneath. Grows in open woods, mixed forests and on streambanks in all prairie provinces. Prefers moist habitat but will tolerate some drought. Also called: soopolallie, russet buffaloberry.

Silver buffaloberry (*S. argentea*)

Soapberry (*S. canadensis*)

Indian Ice Cream

Makes approximately 6 cups

Even with sugar this treat will have a slightly bitter taste, but many people quickly grow to like it.

1 cup soapberries • 1 cup water
4 Tbsp granulated white sugar

Put berries and water into a wide-topped ceramic or glass mixing bowl. *Do not use a plastic bowl or utensils, and make sure that nothing is greasy, or the berries will not whip properly.* Whip the mixture with an electric eggbeater or hand whisk until it reaches the consistency of beaten egg whites. Gradually add the sugar to the pink foam, but not too fast or the foam will "sink." Serve immediately.

Silverberry *Elaeagnus commutata*
Also called: wolf willow

Silverberry (*E. commutata*)

The berries (known as "silverberries") are very dry and astringent, but some northern tribes gathered them for food. Most groups considered the mealy berries famine food. Silverberries were sometimes peeled before being eaten raw or cooked in soup. They were also cooked with animal blood, mixed with lard, or fried in moose fat. For winter storage the berries were mixed with grease and stored for use in soups and broths. They make good jams and jellies and have been used to make wine. The berries are much sweeter after exposure to freezing temperatures.

Some First Nations used the bark as a medicine. A strong decoction was mixed with grease or lotion to take away the sting of frostbite or sunburn. When western meadow rue (*Thalictrum occidentalis*) was added to the mix, it was used to treat hemorrhoids. When sumac roots were added, the decoction was used to treat syphilis. A weak tea was taken as a remedy for chest colds.

Several tribes used the nutlets inside the berries as decorative beads. The fruits were boiled to remove the flesh, and while the seeds were still soft, a hole was made through each. They were then threaded, dried, oiled and polished.

Silverberry flowers can be detected from metres away by their sweet, heavy perfume. Some people enjoy this fragrance, but others find it

TREES & SHRUBS

DESCRIPTION: Thicket-forming, rhizomatous shrub 1–4 m tall, with 2–6 cm long, alternate, lance-shaped leaves covered in dense, tiny, star-shaped hairs (appearing silvery). Flowers strongly sweet-scented, yellow inside and silvery outside, 12–16 mm long, borne in clusters of 1–4 in leaf axils. Grows on well-drained, often calcareous slopes, gravel bars, streambanks and forest edges at low to montane elevations across the prairie provinces.

overwhelming and nauseating. If green silverberry wood is burned in a fire, it gives off a strong smell of human excrement. Some practical jokers enjoy sneaking branches into the fire and watching the reactions of fellow campers.

EDIBILITY: edible, not palatable

FRUIT: Fruits silvery, mealy, about 1 cm long, drupe-like, with a single large nutlet.

SEASON: Flowers June to July. Fruits ripen in September.

Hawthorns *Crataegus* spp.

Red hawthorn (*C. columbiana*)

The fruits, or haws, of this species are edible. Their taste, however, can vary greatly depending on the species, particular tree, time of year and growing conditions. The haws are usually rather seedy, with the flavour described as a range of sweet, mealy, insipid, bitter, astringent or even tasteless. Frosts are known to increase the sweetness of this fruit so it is perhaps better to gather these fruit later in fall.

Historically, these berries were eaten fresh from the tree, or dried for winter use. They were also often an addition to pemmican. The cooked, mashed pulp (with the seeds removed) was dried and stored in cakes as a berry-bread, which could be added to soup or eaten with deer fat or marrow. The Blackfoot made offerings to the hawthorn tree before harvesting the berries, to prevent the fruit from causing stomach cramps when eaten. Haws are rich in pectin, and if boiled with sugar, can be a useful aid in getting jams and jellies to set in lieu of a commercial pectin product. They

Red hawthorn (*C. columbiana*)

Black hawthorn (*C. douglasii*)

can also be steeped to make a pleasing tea or cold drink.

Hawthorn flowers and fruits are commonly used in herbal medicine as heart tonics, though not all species are equally effective. Studies have supported the use of hawthorn extracts as a treatment for high blood pressure associated with a weak heart, angina pectoris (recurrent pain in the chest and left arm owing to a sudden lack of blood in the heart muscle) and arteriosclerosis (loss of elasticity and thickening of the artery walls). Hawthorn is believed to slow the heart rate and reduce blood pressure by dilating the large arteries supplying blood to the heart and by acting as a mild heart stimulant. However, hawthorn has a gradual, mild effect and must be taken for extended periods to produce noticeable results.

Hawthorn tea has also been used to treat kidney disease and nervous conditions such as insomnia. Dark-coloured haws are especially high in flavonoids and have been steeped in hot water to make teas for strengthening connective tissues damaged by inflammation. The haws were sometimes eaten in moderate amounts to relieve diarrhea (some indigenous peoples considered them very constipating). The scientific name *Crataegus* derives from the Greek *kratos*, which means "strength" and refers to the hard quality and durability of the wood. The common name "hawthorn" derives from the Old English word for a hedge, or "haw," and the species was historically planted and worked into hedgerows where its spiky thorns, branching nature and durable wood make it a formidable and lasting barrier, even today.

Hawthorns

Fireberry hawthorn (C. chrysocarpa)

Red hawthorn (*C. columbiana*) grows to 6 m tall, with thorns 4–7 cm long. Dark red-coloured haws are egg-shaped. Leaves distinctly lobed. Grows in open prairies, meadows, streambanks and forest edges in steppe and montane zones in AB.

Fleshy hawthorn (C. succulenta)

EDIBILITY: edible

FRUIT: Fruits are haws, hanging in bunches; small, pulpy, red to purplish pomes (tiny apples) containing 1–5 nutlets.

SEASON: Flowers May to June. Haws ripen late August to September.

DESCRIPTION: Deciduous shrubs or small trees growing 6–11 m tall with strong, straight thorns growing directly from younger branches. Leaves alternate, generally oval, with a wedge-shaped base. Flowers whitish, 5-petalled, 12 mm across, in showy, flat-topped clusters of 6–15, sometimes unpleasant-smelling.

Fireberry hawthorn (*C. chrysocarpa*) is a shrub or small tree, to 6 m tall, with a crooked trunk. Stout branches usually have numerous shiny black thorns, 2–6 cm long. Leaves dull yellowish green, toothed, lobed. Flowers small, 1–1.5 cm wide, white. Haws usually deep red and hairy, often persisting through winter. Found on open, gravelly sites near water throughout the southern regions of the prairie provinces.

Fireberry hawthorn (C. chrysocarpa)

TREES & SHRUBS

Black hawthorn (*C. douglasii*) grows to 11 m tall, with thorns 1–2 cm long. Leaves are dark green, toothed to shallowly lobed. Haws are 1 cm long, purplish black in colour. Grows in forest edges, thickets, streamsides and roadsides in lowland to montane zones throughout the prairie provinces. Also called: thorn apple.

Fleshy hawthorn (*C. succulenta*) is a shrub or multistemmed, shrubby tree, to 8 m tall. Thorns 3–4.5 cm long. Haws bright red, 7 mm–1.2 cm long. Grows in thickets, pastures and woodland edges from southern SK to MB.

Black hawthorn (*C. douglasii*)

Fleshy hawthorn (*C. succulenta*)

INTERESTING: *Hawthorns aren't just important to people, they are also critical to the unique eating habits of the shrike. These birds feed on small birds and rodents, but they don't have talons to hold their prey to eat it. The shrike relies on bushes like hawthorn, which have thorns that they can impale their prey on. Using the thorns, they can tear apart their prey and also store anything they don't eat for later, safely out of reach of other animals.*

Black hawthorn (*C. douglasii*)

Mountain Ashes *Sorbus* spp.

European mountain ash (*S. aucuparia*)

The bitter-tasting fruits of western mountain ash are high in vitamin C and can be eaten raw, cooked or dried; however, most Canadian indigenous groups considered them inedible. After picking, these berries were sometimes stored fresh underground for later use. They were also added to other more popular berries or used to marinate meat such as marmot or as a flavouring for salmon head soup. The green berries are too bitter to eat, but the ripe fruit, mellowed by repeated freezing, is said to be tasty enough. This species has been used to make jams, jellies, pies, ale and also bittersweet wine, and the fruit is also enjoyed cooked and sweetened. In northern Europe, the berries, which can be quite mealy, were historically dried and ground into flour that was fermented and used to make a strong liquor.

A tea made of the berries is astringent and has been used as a gargle for relieving sore throats and tonsillitis. European mountain-ash fruit has been used medicinally to make teas for treating indigestion, hemorrhoids, diarrhea and problems with the urinary tract, gallbladder and heart. Some indigenous peoples rubbed the berries into their scalps to kill lice and treat dandruff.

European mountain ash is a popular ornamental tree, and the native mountain ashes make attractive garden shrubs. These trees are easily propagated from seed sown in autumn. The scarlet fruit can persist throughout winter, and the bright clusters of fruit attract many birds.

EDIBILITY: edible with caution

FRUIT: Berry-like pomes, about 1 cm long, hanging in clusters.

SEASON: Flowers June to July. Fruits ripen August to September.

DESCRIPTION: Clumped, deciduous shrubs or trees with pinnately divided, sharply toothed leaves. Bark smooth, brownish, with numerous lenticels (raised ridges that are actually breathing pores) on young bark, turning grey and rough with age. Leaves compound leaflets, alternate, 9–17 on each stem. Leaflets serrated, narrow, darker above, paler below. Flowers white, about 1 cm across, 5-petalled, forming flat-topped clusters, 9–15 cm wide, strong-smelling. Grows in sun-dappled woods, rocky ridges and forest edges, preferring moist areas and partial to full sun.

European mountain ash (*S. aucuparia*) is an ornamental tree, to 15 m tall, with hairy white buds, leaf stems and leaves (underneath at least), and orange to red fruit. This Eurasian species is widely cultivated and just as widely escaped in AB and SK. Also called: Rowan tree.

Western mountain ash (*S. scopulina*) is a tall shrub, to 4 m tall, with sticky twigs and buds. Shiny green, pointed leaflets are 2–9 cm long, with teeth almost to the base. Orange fruit. Grows in moist to wet open forests, glades and from streambanks to higher elevations in AB and SK. Also called: Greene's mountain ash.

Western mountain ash (*S. scopulina*)

Western mountain ash (*S. scopulina*)

European mountain ash (*S. aucuparia*)

INTERESTING: *European mountain ash fruits contain parasorbic acid and the seeds contain cyanogenic glycoside, both of which are toxic. The parasorbic acid is neutralized with cooking, so the fruit should not be eaten raw, and the seed should be removed before eating.*

Wild Roses *Rosa* spp.

Prairie rose (*R. woodsii*)

Most parts of rose shrubs are edible and the fruit (hips), which remain on the branches throughout winter, are available when most other species have finished for the season. The hips can be eaten fresh or dried and are most commonly used in teas, jams, jellies, syrups and wine. Only the fleshy outer layer of the hips should be eaten (see Warning). Because the hips are seedy, some indigenous peoples considered them famine food rather than regular fare. Some tribes liked to eat them fried with sugar, added to soups or stews, or mixed with pemmican.

Rose petals have a delicate rose flavour with a hint of sweetness and may be eaten alone as a trail nibble, added to teas, jellies and wines or candied. Adding a few rose petals to a regular salad instantly turns it into a gourmet conversation piece, and guests are often surprised at how delicate and sweetly delicious the petals taste. Do not add commercial rose petals to salads, however, as they are often sprayed with chemicals.

Rose hips are rich in vitamins A, B, E and K and are one of our best native sources of vitamin C—three hips can contain as much as a whole orange! They were considered so nutritious that a tablespoon of rosehip jelly was taken every day for good health, as you would with cod liver oil. During World War II, when oranges could not be imported, British and Scandinavian people collected hundreds of tonnes of rose hips to make a nutritional syrup. The vitamin C content of fresh hips varies greatly, but that of commercial "natural" rose hip products can fluctuate even more, as a result of different kinds of processing.

Rose petals have been taken to relieve colic, heartburn and headaches. They were also used to treat sore eyes; boiling water was poured over the petals and the cooled liquid was used as an eye wash. The petals were also ground and mixed with grease to make a salve for mouth sores or mixed with wine to make a medicine for relieving earaches, toothaches and uterine cramps. They were eaten raw to prevent or treat colds and fevers, and as a heart tonic. Rose petals were also chewed and applied as a poultice to treat bee stings. The hips were used to relieve itch and to treat eye problems.

Dried rose petals have a lovely fragrance and are a common ingredient in herbal teas and potpourri. Rose sprigs were traditionally hung on cradle boards to keep ghosts away from babies, and on the walls of haunted houses and in graves to prevent the dead from howling. The hips were used as beads and made into necklaces or toy pipes for children (a hollowed green rose hip was used as the bowl, and a hollow piece of grass, fireweed or willow twig as the pipe stem).

Some native roses can hybridize with each other, resulting in offspring that have mixed traits and can be challenging to positively identify.

EDIBILITY: edible

FRUIT: Fruits scarlet to purplish, round to pear-shaped, berry-like hips, 1.5–3 cm long, with a fleshy outer layer enclosing many stiff-hairy achenes.

SEASON: Flowers June to August. Hips ripen August to September.

DESCRIPTION: Thorny to prickly, deciduous shrubs, often spindly, 0.3–4 m tall. Spines generally straight (introduced rose species, of which there are a few naturalized in Canada, tend to have curved spines). Leaves alternate, pinnately divided into about 5–7 oblong, toothed leaflets, generally odd in number. Flowers light pink to deep rose, 5-petalled, fragrant, usually

Prickly rose (*R. acicularis*)

Wild Roses

growing at the tips of branches. Inhabits a wide range of habitat from dry rocky slopes, forest edges, woodlands and clearings, to roadsides and streamsides at mid- to low-level elevations.

Prickly rose (*R. acicularis*) grows to 1.5 m tall, with bristly, prickly branches and small clusters of deep rose flowers 5–7 cm wide. Hips 1–2 cm long, pear-shaped. Thorns 3–4 mm long. Leaves stalked, 3–9 leaflets (oval in shape, 12–50 mm long), stipules with soft, glandular hairs. Grows in open woods, thickets and edges of fields, streams and pastures, and on rocky slopes in all prairie provinces. Prickly rose is Alberta's floral emblem. Also called: wild rose • *R. bourgeauiana*.

Prickly rose (*R. acicularis*)

Arkansas rose (*R. arkansana*) is a small shrub, 20–50 cm tall, with dense bristly stems that die back near to the base each fall. Prickles of various sizes but generally 3 mm long. Bark reddish brown. Flowers, pink, 3–7 cm wide, clustered at end of branches. Apple-like hips 8–13 mm across. Grows in dry grassy slopes, prairies, banks and open forests in all prairie provinces. Also called: prairie rose.

Arkansas rose (*R. arkansana*)

Arkansas rose (*R. arkansana*)

WARNING: *The dry inner "seeds" (achenes) of the hips are not palatable, and their fibreglass-like hairs can irritate the digestive tract and cause "itchy bum" if ingested. As children, we used to make a great old-fashioned itching powder by slicing a ripe hip in half and scraping out the seeds with these attached hairs. Spread this material to dry, then swirl it in a bowl, and the seeds will drop to the bottom. Skim off the fine, dry hairs—this is your itching powder, guaranteed to work. While all members of the rose family have cyanide-like compounds in their seeds, these are destroyed by drying or cooking.*

TREES & SHRUBS

Prairie rose (*R. woodsii*) is a small shrub, 20–50 cm tall, with well-developed thorns at its joints, no small bristles or prickles on upper stems, small clusters of 3–5 cm-wide flowers. Leaves oval, 5–9 leaflets. Hips globe-shaped 10–20 mm across. Grows in thickets, prairies and on riverbanks in all prairie provinces. Also called: Woods' rose • *R. alcea, R. suffulta*.

Prairie rose (*R. woodsii*)

Rosehip Jelly

Makes 8 x 1 cup jars

2 lbs whole rosehips • 2 lbs apples • 5 cups water • juice of 1 lemon
6 to 8 cloves • small cinnamon stick • white sugar

Carefully wash rosehips and apples. *Slightly unripe apples work best for this recipe as they have a higher pectin content than ripe fruit does.* Core apples and chop roughly. Place the fruit in separate cooking pans with 2½ cups of water in each pan. Add lemon juice, cloves and cinnamon to the pan containing the rosehips. Bring both pans gently to a boil, then reduce heat and simmer until the fruit is soft and pulpy. Place the contents of both pans together in a jellybag and allow the juice to strain through overnight into a clean bowl. *If you want a perfectly clear jelly, do not press or squeeze the bag.*

In the morning, measure the strained liquid and allow for 2 cups of sugar to every 2½ cups of juice. Place the juice and sugar in a thick-bottomed cooking pan. *A thick-bottomed pan is important, because a thin-bottomed pan will get too hot and scald the jelly.* Bring to the boil, stirring and being careful to scrape the bottom of the pan, until the sugar is dissolved. Boil until setting point is reached (when you take some of the liquid on a wide-lipped spoon, blow on it to cool, then start to pour it off the side of the spoon and it gels together). Meanwhile, prepare 8 x 1 cup jars and lids (wash and sterilize jars and lids, and fill jars with boiling water; drain just before use).

Pour the hot jelly into clean, hot, sterilized jars. Seal the jars and place out of the sun to cool.

Plums *Prunus* spp.

American plum (*P. americana*)

Plums make excellent jams, pies, compotes and jellies. The fruit is juicy and delicious to eat fresh from the tree, although the flesh closest to the pit tends to have a sour taste, and the skins can be a bit puckery. American plum and Canada plum are the only native plums in Canada and have a long history of use by First Nations and early settlers throughout the fruit's native range. It is reported that the fruit tastes sweeter if left on the tree until after the first frost. Plums were traditionally eaten fresh, cooked into a sauce, or dried halved, whole or in cakes then reconstituted in winter.

The fruits were crushed and salt was added to treat mouth diseases. One indigenous group used the plum as a seasonal indicator; when the plums were in bloom, they knew it was time to plant their corn, beans and squash.

Edibility: highly edible

Fruit: Juicy, red or orange-coloured plum with a flattened hard pit in the centre, approximately 2.5 cm in size, hanging singly or in clusters of 2–5.

Season: Flowers May to early June. Fruits ripen August to September.

Description: Deciduous shrubs or small trees to 9 m tall, often growing in a crooked, straggling habit and forming thickets. Crown spreading, branches stiff. Bark of young branches covered in horizontal lenticels. Leaves smooth, dark green, paler beneath, alternate, oval, simple, 6–10 cm long. Flowers white, appearing before or with the leaves in early spring, 2–5 per cluster, 5-petalled. Grows in deciduous woodlands, hedges, field margins and along watercourses, preferring edge habitat with good sunlight.

American plum (*P. americana*) has reddish-brown to dark grey bark, older bark rough-textured. Leaves sharp-toothed, narrower and broadest below the middle of the blade, tapering down to a sharp tip compared to Canada plum. The fruit, which is generally considered larger and sweeter than that of Canada plum, often has a slight vertical cleft and a whitish waxy bloom. Grows in prairies, woodlands, pastures, and along roadsides and riverbanks in SK and MB. Also called: wild plum.

Canada plum (*P. nigra*) is more shade tolerant than American plum and grows to 3 m tall. Bark black. Leaves with blunt teeth, doubly toothed along the margins, wider and broadest above the middle of the blade compared to American plum. Flowers sometimes turn pink as they mature. Fruit is shiny, with no whitish bloom, oranged-red on the outside with yellow interior flesh. Grows eastward from southwestern MB but often found outside of this range as the result of ornamental plantings and naturalization. Also called: wild plum.

American plum (*P. americana*)

WARNING: *The leaves, stems and pits of plums contain hydrocyanic acid, a cyanide-producing compound. However, drying, freezing or cooking the fruit eliminates this acid.*

American plum (*P. americana*)

Canada plum (*P. nigra*)

Red Cherries *Prunus* spp.

Pin cherry (*P. pensylvanica*)

These cherries may be eaten raw as a tart nibble, but the cooked or dried fruit is much sweeter and additional sugar further improves the flavour. The fruit can be cooked in pies, muffins, pancakes and other baking, or strained and made into jelly, syrup, juice, sauce or wine. It seldom contains enough natural pectin to make a firm jelly, however, so pectin must be added (see "Hawthorns" for a natural alternative to store-bought preparations). Although wild cherries are small compared to domestic varieties, they can be collected in large quantities. Pitting such small fruits is a tedious job, though, especially because they are too tiny to use with a cherry-pitting tool.

Traditionally, pin cherry fruit were eaten fresh, cooked or dried and then powdered to store for winter use. They were also mixed with bear grease and powdered meat to make pemmican. The berries can be boiled to make a red dye.

When in flower, the pin cherry tree is a dramatic and sweet-scented pleasure to behold so is well worth considering for the ornamental garden. Wild cherry trees are a good addition to garden landscaping as these are a favourite food for many mammals such as chipmunks, rabbits, mice, deer, elk and moose, and birds such as robins and grouse.

EDIBILITY: highly edible

FRUIT: Fleshy drupes (cherries) with large stones (pits), ranging in colour from red to blackish purple to black.

SEASON: Flowers April to June. Fruits ripen July to August.

DESCRIPTION: Deciduous shrubs or small trees growing 1–8 m tall. Trunk and branches reddish brown, often shiny, with raised horizontal pores

TREES & SHRUBS

(lenticels) prominent in stripes on the trunk and larger branches. Leaves smooth, finely toothed, sharp-tipped, 3–10 cm long, alternate. Flowers white or pinkish, about 1 cm across, 5-petalled, forming small, flat-topped clusters, appearing the same time the leaves come out.

Pin cherry (*P. pensylvanica*) grows to 8 m tall. Leaves, slender with long-tapering points, lance-shaped, sharp-toothed, to 10 cm across, two small glands on stalk near base of blade. Flowers in clusters of 5–7 along twigs. Fruit bright red cherries, 4–8 mm long, thin sour flesh, growing in elongated clusters, 10+ per bunch. Grows in moist thickets, woods, riverbanks, forest edges, disturbed areas and clearings throughout the prairie provinces. Also called: bird cherry, fire cherry, Pennsylvania cherry.

Sand cherry (*P. pumila*) is a low, slender shrub, to 1 m tall. Leaves fine-toothed, lance-shaped. Fruit 13–15 mm long, dark purple when ripe. Grows in sandy spots (dunes, beaches) and open areas in calcareous soils from southern SK to MB.

Pin cherry (*P. pensylvanica*)

Sand cherry (*P. pumila*)

WARNING: *Cherry leaves, bark, wood and seeds (stones) contain hydrocyanic acid and therefore can cause cyanide poisoning. The flesh of the cherry is the only edible part. The stone should always be discarded, but cooking or drying destroys the toxins. Cherry leaves and twigs can be poisonous to browsing animals.*

Pin cherry (*P. pensylvanica*)

Pin cherry (*P. pensylvanica*)

Chokecherry *Prunus virginiana*
Also called: wild cherry

Chokecherry (*P. virginiana*)

Chokecherries were among the most important and widely used berries by First Nations in the prairies and indeed across Canada. These fruits were collected after a frost (which makes them much sweeter) and were dried or cooked, often as an addition to pemmican, cooked meat, soups or stews. They were also mixed with grease, fish eggs, or crushed and cooked with lard and sugar or cream. Large quantities were gathered, pulverized with rocks, formed into little cakes about 15 cm in diameter and 2 cm thick and dried for winter use. Another method of storing was to grease and dry them in the sun.

Today, chokecherries are used to make beautiful-coloured jellies, syrups, sauces and beer as well as wine. The raw cherries are sour and astringent, particularly if they are not fully ripe, so they cause a puckering or choking sensation when they are eaten—hence the common name "chokecherry." One unimpressed early European traveller in 1634 is reported to have written that "chokecherries so furre the mouthe that the tongue will cleave the roofe, and the throat wax hoarse"! After chokecherries have been cooked or dried, however, they are much sweeter and lose their astringency.

Dried, powdered cherry flesh was eaten to improve appetite and relieve diarrhea and bloody discharge of the bowels. Chokecherry juice was taken to treat diarrhea, sore throats and postpartum hemorrhage.

EDIBILITY: highly edible (when fully ripe, after a frost or sweetened)

TREES & SHRUBS

Fruit: Red, black to mahogany-coloured, shiny, around 8 mm across, growing in heavy and generous trusses. Some reports indicate that the red fruit have a nicer flavour than the darker-coloured ones.

Season: Flowers May to June. Fruits ripen August to September.

Description: Deciduous shrub or most often small tree growing to 6 m tall. Bark smooth, greyish, marked with small horizontal slits (slightly raised pores called lenticels). Leaves alternate, 3–10 cm long, broadly oval, finely sharp-toothed, bright green above and paler below, with 2–3 prominent glands near the stalk tip. Flowers creamy white, 10–12 mm across, 5-petalled, forming bottlebrush-like clusters 5–15 cm long, at the ends of branches. Dark red to black cherries hang in clusters. Chokecherry grows in deciduous woods, open sites, streams and forest edges throughout the prairie provinces.

Warning: *Like other species of* Prunus *and* Pyrus, *all parts of the chokecherry (except the flesh of the fruit) contain cyanide-producing glycosides. There are reports of children dying after eating large amounts of fresh chokecherries without removing the stones. Cooking or drying the seeds, however, appears to destroy most of the glycosides. Chokecherry leaves and twigs are poisonous to animals.*

Saskatoon *Amelanchier alnifolia*

Also called: serviceberry, Canada serviceberry, juneberry, shadbush • *A. florida*

Saskatoon (*A. alnifolia*)

These sweet fruits were and still are extremely important to many indigenous peoples across Canada; they were one of, if not the most important fruit for many indigenous groups in the prairie provinces. Indeed, there is a well-documented history of extensive landscape management through fire, weeding and pruning to encourage the healthy growth of this important species.

The Canadian name for this fruit, saskatoon, is derived from the Cree name for these berries, "misaskatomina." Saskatoons were eaten fresh, alone or with oil, or often mixed with less palatable berries as a sweetener. The Blackfoot made sausages with the berries and fat, and the Assiniboine mixed them with dried prairie turnips (*Psoralea esculenta*). When the berries were dried, they were later eaten in that state, or rehydrated by boiling and added to soups or stews. Dried saskatoons were the principal berries mixed with meat and fat to make pemmican.

Large quantities of the berries were harvested and stored for consumption during winter. Two common ways they were processed for storage were either sun-dried whole like raisins, or cooked, mashed and dried into cakes or loaves. Indigenous groups in the

TREES & SHRUBS

prairies used saskatoons as an important trade item with early explorers and fur traders.

Today, the berries are commonly used much like blueberries in pies, pancakes, puddings, muffins, jams, jellies, sauces, syrups and wine.

Historically, saskatoon juice was taken to relieve stomach upset and was also boiled to make drops to treat earache; green or dried berries were used to make eye drops. The fruits were given to mothers after childbirth for afterpains and were also prescribed as a blood remedy. The berry juice, which easily stains your hands when picking, makes a good purple-coloured dye.

Saskatoons are excellent ornamental, culinary and wildlife shrubs. They are hardy and easily propagated, with beautiful white blossoms in spring, delicious fruit in summer and colourful, often scarlet leaves in autumn. Many improved garden cultivars are readily available for superior fruit production in the home garden.

EDIBILITY: highly edible

FRUIT: Juicy, berry-like pomes, red at first, ripening to purple or black, sometimes with a whitish bloom, 6–12 mm across.

SEASON: Flowers May to June. Fruit ripens July to August.

DESCRIPTION: Shrub or small tree, to 5 m tall, often forming thickets. Bark is smooth, grey to reddish brown. Leaves alternate, coarsely toothed on the upper half, leaf blades 2–5 cm long, oval to nearly round, blue-green turning yellowish-orange to reddish-brown in autumn. Flowers white, 9–12 mm across, forming short, leafy clusters near the branch tips. Grows at low to middle elevations in prairies, thickets, hillsides and dry, rocky shorelines, meadows, open woods throughout the prairie provinces.

Saskatoon

Pemmican

Makes 6 cups

Pemmican uses the same drying temperature as fruit leather (see p. 93), so make both recipes at the same time!

3 Tbsp salted butter
3 Tbsp brown sugar
¼ tsp dried powdered ginger
¼ tsp ground cloves
¼ tsp ground cinnamon
4 cups saskatoons or blueberries
4 cups beef jerky, chopped into small pieces
½ cup chopped almonds, walnuts or hazelnuts (optional)
½ cup sunflower seeds (optional)

Gently heat butter with sugar and spices in a heavy-bottomed pot. Mash berries and add to pot. Simmer, stirring constantly, for about 5 minutes. Let mixture cool, then mix in jerky and nuts and/or seeds. Grease a rimmed baking sheet, spread mixture evenly on sheet and dry overnight in oven at 150° F.

Saskatoon Squares

Makes 12 squares

¼ cup butter
⅔ cup brown sugar
1 Tbsp vanilla
1 large egg, beaten
1 cup flour
1 tsp baking powder
½ tsp salt
½ tsp cinammon
½ cup frozen saskatoons (or wild blueberries)
½ cup chopped walnuts or almonds

Melt butter gently in a pot, then remove from heat and stir in sugar, vanilla and beaten egg. Mix dry ingredients in a bowl. Make a shallow depression in the middle, and gradually mix in wet ingredients from the saucepan. When well mixed, add saskatoons and nuts. Pour into an 8-inch pan and bake at 350° F for 35 minutes. Remove from oven and cool before cutting into squares.

Saskatoon Crumble

3–4 cups saskatoons
¼ cup white sugar
1 cup rolled oats
½ cup flour
¾ cup brown sugar
½ cup cold butter

Add saskatoons to a 8 x 8-inch baking dish. Sprinkle with white sugar. Combine remaining ingredients in a bowl to make a crumbly mixture. Sprinkle over the saskatoons. Bake at 350° F for about 30 minutes. Enjoy!

TREES & SHRUBS

Currants *Ribes* spp.

Prickly currant (*R. lacustre*)

Currants are common and widespread throughout Canada and were eaten by many First Nations. All are considered edible, but some are tastier than others, and some (such as sticky currant) are considered emetic in large quantities and are best avoided. These fruit are high in pectin and make excellent jams, jellies and syrup either alone or mixed with other fruit (such as high bush cranberries), and are delicious with meat, fish, bannock or toast. They are also good in pies, cakes, muffins, bannock and breads. Currants have historically also been mixed with other berries and used to flavour liqueurs or fermented to make delicious wines, but raw currants tend to be very tart.

Currants were traditionally eaten fresh, cooked or pounded into cakes and dried for winter use by many indigenous peoples. They were sometimes added to pemmican. A favourite Ojibway dish in the winter was wild currants cooked with sweet corn. Currants were also boiled to make tea.

While skunk currants are said to have an unpleasant odour and taste, multiple First Nations such as the Woodlands Cree ate them, often in considerable quantity. They are delicious cooked.

Some Native peoples considered the spiny branches (and by extension, the fruit) of prickly currant to be poisonous, while other groups regularly consumed the berries, such as the Stoney of Alberta. Golden currant is one of the most flavourful and pleasant-tasting currants.

In Europe, currant juice is taken as a natural remedy for arthritic pain. Black currant seeds contain gamma-linoleic acid, a fatty acid that has been used in the treatment of migraine headaches, menstrual problems, diabetes, alcoholism, arthritis and eczema. A decoction of northern black currant leaves and berries was sometimes taken for general sickness. Some tribes boiled the berries in water and wrapped them in a poultice to treat sore eyes. Some Native peoples believed that northern black currant had a calming effect on children, so sprigs were often hung on baby carriers. For some tribes, currant shrubs growing by lakes were an indicator of fish, and in some legends, when currants dropped into the water, they were transformed into fish. The Woods Cree did not eat the berries and called them athikimin, or "frog berry."

The name "currant" comes from the ancient Greek city of Corinthe, where a small purple grape (*Uva corinthiaca*) was historically grown and sold commercially as a "currant." For more information on closely related species, see Gooseberries.

EDIBILITY: edible

FRUIT: Fruit colour varies from bright red to green to black, as do the size, sweetness and juiciness, depending on the species and individual location.

SEASON: Flowers April to July. Fruits ripen July to August.

DESCRIPTION: Erect to ascending, deciduous shrubs, 0.5–3 m tall. Leaves alternate, 3- to 5-lobed, usually rather maple leaf–like, sometimes dotted with yellow, crystalline resin glands. Flowers small (about 5–10 mm across), with 5 petals and 5 sepals, borne in elongating clusters in spring. Fruits tart, juicy berries (currants), often speckled with yellow resinous dots or bristling with stalked glands.

Black currant (*R. americanum*) is a small, non-prickly shrub usually growing 90–120 cm in height. Leaves simple, rounded and alternate, to 3–8 cm long, 10 cm wide, palmately lobed, with 3–5 pointed lobes with doubly-toothed edges. The surfaces of the leaves are scattered with resinous dots. Flowers creamy white to yellowish, bell-shaped and hanging in

Black currant (*R. americanum*)

Currants

clusters from the leaf axils. Fruit a black berry, globular, smooth, each with a characteristic residual flower at the end. Found in damp soil along streams, wooded slopes, open meadows and rocky ground from southern AB to MB.

Golden currant (*R. aureum*) grows 1–3 m tall and is named for its showy, bright yellow flowers, not its smooth fruits, which range from black to red and sometimes yellow. No spines or prickles. Leaves have 3 widely spreading lobes and few or no glands. Inhabits streambanks and wet grasslands to dry prairies and open or wooded slopes from southern AB to southwestern SK.

Skunk currant (*R. glandulosum*) is a loosely branched, unarmed shrub 0.5–1 m tall with spreading stems. Bark is brownish. Leaves 3–8 cm wide, 5–7 lobed, glabrous, with strong

Skunk currant (*R. glandulosum*)

Golden currant (*R. aureum*)

skunk-like odour when bruised. Flowers white or reddish, stalks not jointed below the flowers, in clusters of 6–15. The hairy (gland-tipped), dark red fruits, 6–8 mm, are nearly round in shape and considered not very nice to eat (unpleasant odour and flavour). Found in swamps, thickets, moist woods, clearings and streambanks across the prairie provinces.

Northern black currant (*R. hudsonianum*) grows to 2 m tall, with no bristles or prickles. Has elongated clusters of 6–12 saucer-shaped, white flowers and shiny, resin-dotted, black berries (5–10 mm across). Leaves are relatively large (5–7 cm) with yellow resin dots on the lower surface. Fruit strong-smelling and often bitter-tasting. Found in wet woods and on rocky slopes across the prairies.

Northern black currant (*R. hudsonianum*)

Prickly currant (*R. lacustre*) is an erect to spreading shrub, 0.5–2 m tall, covered with numerous small, sharp prickles, with larger, thick thorns at leaf nodes. Bark on older stems is cinnamon coloured. Branches spiny, prickly. Leaves 3–4 cm across, hairless to slightly hairy. Flowers reddish to maroon, saucer-shaped, about 6 mm wide, in hanging clusters of 7–15. Berries dark purple or black, 5–8 mm wide, bristly with glandular hairs. Found in moist woods and streambanks to drier forested slopes and subalpine ridges across all prairie provinces. Also called: bristly black currant, swamp gooseberry.

Trailing black currant (*R. laxiflorum*) is a trailing, spreading plant, occasionally vining, with branches growing along the ground, no prickles or

Prickly currant (*R. lacustre*)

WARNING: *Sticky currant is reported to cause vomiting, even in small quantities.*

spines, and usually less than 1 m tall. Flowers greenish white to reddish purple. Fruits purplish black, stalked glandular hairs, waxy bloom. Inhabits clearings, disturbed sites such as avalanche tracks and roadsides and moist forests at mid to low elevations in AB.

Red swamp currant (*R. triste*) is an unarmed, reclining to ascending shrub, to 1 m tall. Flowers reddish or greenish purple, small, several (6–15) in drooping clusters; flower stalks jointed, often hairy and glandular. Fruits bright red, smooth and sour but palatable. Found in moist coniferous forests, swamps, on streambanks and montane, rocky slopes across the prairies. Also called: northern red currant • *R. propinquum*.

Sticky currant (*R. viscosissimum*) is an unarmed, loosely branched shrub, to 2 m tall, with erect to spreading stems. Leaves heart-shaped at the base, 2–10 cm wide and glandular-sticky.

Red swamp currant (*R. triste*)

Red swamp currant (*R. triste*)

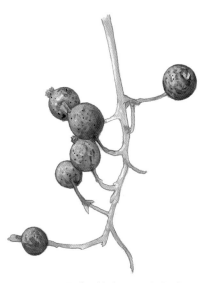

Trailing black currant (*R. laxiflorum*)

TREES & SHRUBS

Flowers white or creamy in clusters of up to 16. Fruits bluish black, hairy, glandular berries. Grows in moist to dry forests and woodlands at montane to subalpine elevations in southern AB.

Prickly currant (*R. lacustre*)

Northern black currant (*R. hudsonianum*)

Sticky currant (*R. viscosissimum*)

Northern black currant (*R. hudsonianum*)

Gooseberries *Ribes* spp.

Northern gooseberry (*R. oxyacanthoides*)

While gooseberries and currants are closely related species, they are generally different: gooseberries have spines or prickles on their stems (currants are not thus "armed"), and gooseberry fruit are usually borne in small clusters or singly (currants are in elongated clusters generally more than 5). However, common names are inconsistent, so some "gooseberries" don't have spines and some "currants" do!

Then there is *Ribes lacustre* (bristly black currant, swamp gooseberry), which has characteristics of both currants (fruit in clusters of 7–15) and gooseberries (has prickles).

All gooseberries are edible raw, cooked or dried, but flavour and sweetness vary greatly with species, habitat and season. All are high in pectin and make excellent jams and jellies, either alone or mixed with other fruits.

Gooseberries can be eaten fresh and are also good in baked goods such as pies. Traditionally, these fruits were eaten with grease or oil, combined with sweet corn or added to soups, and also mashed (usually in a mixture with other berries) and formed into cakes that were dried and stored for winter use. Dried gooseberries were sometimes included in pemmican, and dried gooseberry and bitterroot cakes were sometimes a trade item. Because of their tart flavour, gooseberries can be used much like cranberries. They make a delicious addition to turkey stuffing, muffins and breads. Timing is important, however, when picking these fruit. Green berries are too sour to eat, and ripe fruit soon drops from the branch. Sometimes green berries can be collected and then stored so that they ripen off the bush.

Too many gooseberries can cause stomach upset, especially in the uninitiated. Because of the large number of species and wide distribution of gooseberries, there is a very large spectrum of uses for this genus. They were commonly eaten or used in teas for treating colds and sore throats, a use that may be related to their high vitamin C content. Teas made from gooseberry leaves and fruits were given to women whose uteruses had slipped out of place after too many pregnancies. Gooseberry tea was also used as a wash for soothing skin irritations such as poison-ivy rashes and erysipelas (a condition with localized inflammation and fever

White-stemmed gooseberry (*R. inerme*)

Gooseberries

Northern gooseberry (*R. oxyacanthoides*)

caused by a *Streptococcus* infection). Gooseberries have strong antiseptic properties and extracts have proved effective against yeast (*Candida*) infections. Picking this fruit can be a formidable task, though, because of the often-thorny stems. Indeed, gooseberry thorns can be so large and strong that they were historically used as needles for probing boils, removing splinters and even applying tattoos! A traditional way to pick gooseberries and avoid the sharp prickles is to lay a tarp on the ground around the bushes and use a stick to hit the bush and knock the ripe berries off the plants.

The common name "gooseberry" comes from an old English tradition of stuffing a roast goose with the berries.

EDIBILITY: edible

FRUIT: Fruits smooth, green to purplish (when ripe) berries, about 1 cm across.

SEASON: Flowers May to June. Fruits ripen July to August.

DESCRIPTION: Erect to sprawling deciduous shrubs with spiny branches. Leaves alternate, maple leaf–like, 3- to 5-lobed, about 2.5–5 cm wide. Flowers whitish to pale greenish yellow, to

1 cm long, tubular, with 5 small, erect petals and 5 larger, spreading sepals, in 1- to 4-flowered inflorescences in leaf axils.

White-stemmed gooseberry (*R. inerme*) is an erect or spreading shrub to 3 m. Stem hairless, bristly, with spines and black resin glands at nodes. Flowers with white to pinkish petals, 1–1.5 mm long. Fruit hairless, to 1 cm, ranging in colour from greenish, purplish to black. Inhabits meadows, woodlands, foothills and montane forests in southwestern AB.

Northern gooseberry (*R. oxyacanthoides*) is an erect to sprawling shrub to 1.5 m tall, branches bristly, often with 1–3 spines to 1 cm long at nodes; internodal bristles absent or few. Twigs grey to straw-coloured, older bark whitish grey. Leaves heart-shaped at base, usually have gland-tipped hairs on bottom. Fruit black when ripe, 4–10 mm across. Inhabits wet forests, thickets, clearings, open woods and exposed rocky sites in all prairie provinces. Also called: smooth gooseberry, Canadian gooseberry • *R. hirtellum, R. setosum.*

Northern gooseberry (*R. oxyacanthoides*)

White-stemmed gooseberry (*R. inerme*)

Raspberries *Rubus* spp.

Arctic dwarf raspberry (*R. arcticus*)

Raspberries and relatives (salmonberry, thimbleberry and cloudberry) are all closely related members of the genus *Rubus*. The name *Rubus* means "red" and refers to the fruit colour of many *Rubus* species. The best way to distinguish these berries from blackberries is by looking at their fruits: if they are hollow like a thimble, they are raspberries (or relatives), and if they have a solid core, they are blackberries.

Wild raspberries are one of our most delicious native berries and are fabulous fresh from the branch or made into pies, cakes, puddings, cobblers, jams, jellies, juices, syrups and wines. Since the cupped fruit clusters drop from the receptacle when ripe, these fruits are soft and easily crush to a juicy mess when gathered.

Raspberries were a popular and valuable food of indigenous peoples and were often gathered and made into dried cakes (e.g., boiled and sun dried in birch bark baskets) either alone or with other berries for winter use. These cakes were reconstituted by boiling, or eaten as an accompaniment

Trailing raspberry (*R. pubescens*)

TREES & SHRUBS

Trailing wild raspberry (*R. pedatus*)

Arctic dwarf raspberry (*R. arcticus*) is a low, herbaceous plant (sometimes woody at the base), to 15 cm tall, with typically 2–4 leaves, each with 3 leaflets, no prickles or bristles and pink to reddish flowers. Fruits deep red to dark purple, in clusters of 20–30 druplets. Found in bogs, wet meadows and tundra in all prairie provinces. Also called: nagoonberry • *R. acaulis*.

Wild red raspberry (*R. idaeus*) is an erect shrub, to 2 m tall, prickly, growing in thickets as it spreads by underground rhizomes. Leaves pinnately divided into 3–5 leaflets. Flowers white, 5-petalled, appearing singly or in small clusters at ends of stems. Fruit bright red, virtually identical to the domesticated raspberry, but smaller. Grows in thickets, open woods, fields and on rocky hillsides across the prairies. Also called: American red raspberry, dwarf nagoonberry • *R. strigosis*.

to dried meat or fish. Fresh or dried leaves of this species have been used to make tea, and the flowers make a pretty addition to salads. Although trailing wild raspberry fruit are delicious and juicy, they were not traditionally gathered in quantity because of their small size and the difficulty in picking them.

Raspberry leaf tea and raspberry juice boiled with sugar have been gargled to treat mouth and throat inflammations. The berries have also been boiled to make red dye.

EDIBILITY: highly edible

FRUIT: Juicy, red to black drupelets aggregated into clusters that fall from the shrub without the fleshy receptacle (raspberries and relatives have a hollow core).

SEASON: Flowers June to July. Berries ripen July and August.

DESCRIPTION: Armed or unarmed, perennial shrubs or herbs, 15 cm–4 m tall. Leaves deciduous, lobed or compound (divided into leaflets). Flowers white to pink.

Wild red raspberry (*R. idaeus*)

87

Raspberries

Trailing wild raspberry (*R. pedatus*) is a trailing herb from long, creeping stems to 1 m long, rooting at the nodes and producing short, erect stems bearing flowers and 1–3 leaflets, usually less than 10 cm tall. Leaflets alternate, 3–5, coarsely toothed, no prickles or bristles. Flowers solitary, on slender stalks, white, petals spreading or bent backwards. Fruits dark or bright red clusters of flavourful and juicy drupelets (raspberries), sometimes only 1 drupelet per fruit. Found in low to subalpine elevations in bog forests, streambanks and moist, mossy forests in AB. Also called: strawberry leaf raspberry, creeping raspberry, five-leaved bramble.

Trailing raspberry (*R. pubescens*) is a slender, trailing, soft-hairy shrub, unarmed, to 30 cm tall and often more than 1 m long (vegetative stems are ascending at first, then reclining), rooting where the nodes touch the ground. Leaves alternate, double toothed, 5–15 cm long, with 3–5 leaflets. Flowers 5-petalled, 2–3 on erect shoots, 6–10 mm long, blossoms

Wild red raspberry (*R. idaeus*)

Wild red raspberry (*R. idaeus*)

Trailing wild raspberry (*R. pedatus*)

white, rarely pinkish. Fruits dark red drupelets, to 1 cm, smooth, several, not easily separating from spongy receptacle. Found on damp slopes, rocky shores, moist woods and low thickets in all prairie provinces. Also called: dewberry, dwarf red blackberry.

Arctic dwarf raspberry (*R. arcticus*)

Prairie Berry Cordial

Makes approximately 4 x 1 cup jars if 8 cups of fruit are used

up to 8 cups of any freshly picked juicy berry (such as raspberry or blueberry)
white vinegar • sugar

Carefully pick through the fruit to remove any debris or insects. Be particularly wary of stink bugs, which are about 1 cm in size, green to brownish in colour, flat-backed with a hard carapace, and emit a rank stench if bitten into: they will ruin the entire batch of cordial!

Place the berries in a large glass jar and crush somewhat firmly with a potato masher. Pour enough white vinegar into the jar to just barely cover the fruit mash (roughly an 8:1 ratio). Stir vigorously, put a firm lid on the jar, then let it sit somewhere warm out of direct sunlight for 1 week, stirring once a day.

After a week, strain the mixture overnight through a jellybag. *Resist squeezing it or you will push solids through the bag, resulting in a cloudy end-product with sediment*. The leftover fruit mash can be used in muffins or pancakes.

Measure out the resulting juice into a thick-bottomed saucepan and add 1 cup white granulated sugar for every 1 cup of juice. Slowly bring to a boil to fully dissolve the sugar. Let cool and place in washed, sterilized Mason-type jars for storage. Other glass containers such as maple syrup bottles with rubber-sealed tops also work well.

To make the cordial, mix the concentrate in a 6:1 ratio with cold water. Garnish with a sprig of fresh mint or some frozen wild berries.

Wild red raspberry (*R. idaeus*)

Wild Berry Dressing

Makes about 2 cups

This dressing keeps well in the fridge for up to 10 days.

1 cup mixed tangy wild berries such as raspberries, thimbleberries or huckleberries
½ cup olive oil • ¼ cup apple cider vinegar
1 tsp sugar • 2 cloves crushed garlic • 1 tsp salt

Crush the berries, then mix with all the remaining ingredients in a small jam jar. Screw on the lid tightly and shake vigorously.

Thimbleberry *Rubus parviflorus*

Thimbleberry (*R. parviflorus*)

Thimbleberry is a delicious native berry and was highly regarded by all First Nations in its range. The fruit is easy to pick as it can grow in large clusters and appears on the plant as a bright red treasure amid soft, maple leaf–like leaves (yay, no sharp prickles or spines!). The taste is somewhat like a raspberry, but more intense and flavourful with a sharper "tang." The ripe fruits are great in pies, jams or tarts. The fruit, which is rather coarse and not overly juicy, dries and keeps well. This species can also be gathered by cutting the stems of the unripe fruit, which will ripen later in storage. The berries, while pink and not fully ripe, make a tasty sweet-and-sour nibble.

Traditionally, these fruit were gathered, mashed either alone or with other seasonal fruits and dried into cakes for winter use or trade. Tender shoots of this plant were harvested and peeled in the early spring as a green vegetable. The berries were used to treat chest disorders, as a dye for tanned robes and were applied to quivers to "strengthen" them.

The large leaves of thimbleberry served many purposes for some indigenous peoples. They were used to whip soapberries; to wipe the slime from fish; to line and cover berry baskets; and as a mat to dry other kinds of berries. If you're out in the woods and have forgotten your toilet paper, thimbleberry leaves are soft and tough and make an excellent substitute.

EDIBILITY: highly edible

FRUIT: Bright red, shallowly domed (like a thimble), raspberry-like hairy drupelets, in clusters held above the leaves.

SEASON: Flowers April to May. Fruits ripen July to August.

DESCRIPTION: Erect shrub, 0.5–3 m tall, main-stemmed, with no prickles or spines, spreading by underground rhizomes and forming dense thickets. Bark light brown and shredding on mature stems, green on newer stems. Leaves large (up to 15 cm), soft, fuzzy, maple leaf–like, palmate, 3–7 lobed, alternate, toothed around margins, fine hairs above and below. Flowers white, 5-petalled, large to 4 cm, long-stemmed in terminal clusters of 3–11. Found in moist open sites such as road edges, shorelines and riverbanks, montane areas and open forests at low- to mid-elevations in southwestern AB.

BERRY BASKET: *Are you out in the woods and have come across an unexpected and beautifully rich patch of wild berries but don't have a container to gather them into? Making a simple temporary basket out of thimbleberry leaves is easy. Pick the largest leaf you can find and then snap off the stem. Fold the outer soft leaf edges together to form a funnel shape (the stem is at the narrow, bottom edge of this funnel, the leaf tips forming the wider top brim), then use the stem to prick through the two leaf folds where they overlap and "sew" the funnel together. If you still have a small hole at the bottom of your funnel, line this with part of another leaf.*

Wild Berry Juice

Makes approximately 4½ cups

3 cups any sweet berries such as blueberries, bilberries or thimbleberries
2 cups water • sugar to taste

Pick over berries to remove any debris and place them in a saucepan with the water. Mash the mixture with a wooden spoon or potato masher, then simmer until berries are soft. Strain the mixture through a jellybag, fine-mesh sieve or cheesecloth, add sugar to taste, then let the juice cool before serving.

Cloudberry *Rubus chamaemorus*

Also called: bake-apple

Cloudberry (*R. chamaemorus*)

Cloudberry was historically, and still is, a principal food for northern indigenous peoples. The juicy berries are delicious, with a distinctive tart taste that some reports say is acquired. Cloudberries have twice as much vitamin C per volume as an orange and were an important food against scurvy for First Nations and early northern immigrants. Traditionally, these summer fruits were in stored in cold water or oil in seal pokes (containers made by cleaning, inflating and drying a complete sealskin), wooden barrels or underground caches, with other berries or with edible greens. In the past, indigenous people ate cloudberries with animal oil and sometimes mixed them with meat or fish. Today, the berries are mixed with cream or milk and sugar, and are made into jams and jellies, pies and other baking.

The Latin name for this species derives from the Greek words *chamai*, meaning "on the ground," and *moros*, meaning "mulberry."

EDIBILITY: highly edible

FRUIT: Raspberry-like in appearance, made up of 5–25 drupelets, red when unripe, amber to yellow when mature.

SEASON: Flowers May to June. Fruits ripen in August.

DESCRIPTION: A low, unbranched herb, to 25 cm tall, with 1–3 leaves per stem. Leaves round to kidney-shaped (not divided into leaflets), shallowly 5- to 7-lobed, no prickles or bristles. Flowers single, white, at end of stem, the male and female flowers on different plants. Found in peat bogs and peaty forests at northern latitudes in all prairie provinces.

Berry Fruit Leather

Makes 1 baking sheet of fruit leather

4 cups crushed berries (all one kind or a mix) • 2 cups apple sauce • ½ cup sugar

Mix the berries and sugar together in a pot on medium heat until the sugar is dissolved. Put the mixture through a food mill to remove any stems or seeds, then add the apple sauce and stir until well mixed. Grease a rimmed baking sheet and pour mixture in. Use a spatula to spread the mixture to an even thickness on the baking sheet, because the fruit leather will not dry evenly otherwise. Place in a food dehydrator or an oven at 150° F until firm to the touch and dry enough to peel off. Remove from the dehydrator or oven and let cool. Use scissors to cut the leather into strips. Cool the strips and store in an airtight container or Ziploc® bag.

Bearberries *Arctostaphylos* spp.

Red bearberry (*A. alpina* var. *rubra*)

Bearberries are rather mealy and tasteless, but they are often abundant and remain on branches all year, so they can provide an important survival food. Many First Nations traditionally ate these berries; they were dried or buried fresh in birch bark containers for winter use. To reduce the dryness, bearberries were often cooked with bear fat or fish eggs, or they were added to soups or stews. Sometimes, boiled berries were preserved in oil and served whipped with snow during winter. Boiled bearberries, sweetened with syrup or sugar and served with cream, reportedly make a tasty dessert. They can also be used in jams, jellies, cobblers and pies, or dried, ground and cooked into a mush. Apparently, if the berries are fried in grease over a slow-burning fire, they eventually pop, rather like popcorn. Scalded mashed berries, soaked in water for an hour or so, produce a spicy, cider-like drink that can be sweetened and fermented to make wine.

Although fairly insipid, juicy alpine bearberries are probably among the most palatable fruits in the genus, but because they grow at high elevations and northern latitudes, they have been the least used. Hikers sometimes chew the berries and leaves to stimulate saliva flow and relieve thirst.

The fruit was mixed with grease to treat children with diarrhea, and the dried berries can be strung into necklaces. Bearberries, as evidenced by their name, are known to be a favourite fruit of bears.

EDIBILITY: edible, not palatable

FRUIT: Small 0.5–1 cm fruits, bright red to purplish black.

SEASON: Flowers May to July. Berries ripen August to September.

WILD GARDENING: *Bearberries are tough plants and are an excellent choice for planting on difficult or unstable slopes, or particularly dry areas of your garden. The berries also provide food for wildlife.*

DESCRIPTION: Evergreen or deciduous shrubs with clusters of nodding, white or pinkish, urn-shaped flowers and juicy to mealy, berry-like drupes containing 5 small nutlets.

Alpine bearberry (*A. alpina*) is a trailing shrub, to 15 cm tall. Leaves thin, wrinkled and veiny, slightly leathery, oval, 1–5 cm long, with hairy margins (at the base) that often turn red in autumn, the previous year's dead leaves are usually evident. Flowers small (4–6 mm long), producing mealy fruits that are shiny and purplish black, 5–10 mm in diameter. Grows in moderately well-drained, rocky, gravelly and sandy soils on tundra, slopes and ridges in northern areas across the prairie provinces. Also called: black alpine bearberry, whortleberry • *Arctuous alpina*. Note that the common name "whortleberry" is also a common name for a totally unrelated species, *Vaccinium myrtillus*.

Common bearberry (*A. uva-ursi*)

Bearberries

Red bearberry (*A. alpina* var. *rubra*)

Red bearberry (*A. alpina* var. *rubra*)

Red bearberry (*A. alpina* var. *rubra*) is similar to alpine bearberry, but it has longer leaves (to 9 cm long) with hairless margins; the leaves of the previous year are not persistent. Flowers 2–4 in terminal cluster, berries bright red. Grows in the same habitats as alpine bearberry and over the same range.

Common bearberry (*A. uva-ursi*) is a trailing, evergreen shrub to 15 cm tall, often forming mats with branches 50–100 cm long. Leaves alternate, leathery, evergreen, spoon-shaped, 1–3 cm long. Small (4–6 mm long) flowers appear from May to June and produce dull red, 5–10 mm diameter, mealy fruits by late summer. Grows in well-drained, often gravelly or sandy soils in open woods and rocky, exposed sites across the prairie provinces. Also called: kinnikinnick, mealberry, sandberry • *Arctuous rubra*.

Alpine bearberry (*A. alpina*)

Common bearberry (*A. uva-ursi*)

TREES & SHRUBS

Alpine bearberry (*A. alpina*)

Common bearberry (*A. uva-ursi*)

Common bearberry (*A. uva-ursi*)

Black Crowberry *Empetrum nigrum*

Also called: moss berry, curlew berry

Black crowberry (*E. nigrum*)

Next to cranberries and blueberries, crowberries are one of the most abundant edible wild fruits found in northern Canada and were a vital addition to the diets of northern First Nations. Because crowberries are almost devoid of natural acids, they can taste a little bland and were often mixed with blueberries or lard or oil and in more modern times with sugar and lemon. Their taste does seem to vary greatly with habitat, and the flavour of the berries has been described within a range of bland, tasting like turpentine, to most delicious! Their taste improves after freezing or cooking, however, and their sweet flavour peaks after a frost.

The fruits are high in vitamin C, about twice that of blueberries, and are also rich in antioxidant anthocyanins (the pigment that gives the berries their black colour). Their high water content was a blessing to hunters seeking to quench their thirst in the waterless high country. As crowberries have a firm, impermeable skin and are not prone to becoming soggy, they are ideal for making muffins, pancakes, pies, jellies (but pectin needs to be added), preserves and the like. A fine dessert is made by cooking the berries with a little lemon juice and serving them with cream and sugar.

Crowberries are usually collected in autumn, but because they often persist on the plant over winter, they can be picked (snow depth permitting) through to spring if the wildlife doesn't get them first. It can take up to 1 hour to pick 2 cups of berries because the fruit is small. Consuming too many berries may cause constipation, so they were historically prescribed for diarrhea. The berries make a reasonable black-coloured dye.

EDIBILITY: highly edible

FRUIT: Black, shiny, berry-like drupes to 9 mm, sometimes overwintering. Contains 6–9 large, inedible seeds.

SEASON: Flowers May to August. Fruits ripen July to November.

DESCRIPTION: Evergreen dwarf or low shrub, 5–15 cm tall, prostrate and mat-forming, to 40 cm long. Leaves dark to yellowy-green to wine-coloured, 2–6 mm long, alternate but growing so closely together as to appear whorled, needle-like, deeply grooved beneath. Flowers inconspicuous, 1–3, pink to purplish, in leaf axils, 3 petals and sepals, petals 3 mm long, with male and female flowers separate but on the same plant. Grows prolifically in bogs, moist shady forests, low-lying headlands, dry, acidic, rocky or gravelly soil on slopes, ridges and seashores in tundra, muskeg and spruce forests at all elevations in the prairie provinces.

False-wintergreens *Gaultheria* spp.

Hairy false-wintergreen (*G. hispidula*)

The small, sweet berries of this species are delicious and can be eaten fresh, frozen for later use, served with cream and sugar or cooked in baking and sauces. Their flavour improves upon freezing, so they are at their best in winter after the first frost (even from under the snow if you are persistent!), or in spring when they are plump and juicy.

The berries were historically mixed with teas and were used to add fragrance and flavour to liqueurs. Occasionally, large quantities were picked and dried like raisins for winter use. The young leaves can be an interesting trailside nibble or added to salads as well as used to make a strong, aromatic tea that is said to make a good digestive tonic. The wintergreen flavour can be drawn out if the bright red leaves are first fermented.

During the American Revolutionary War, wintergreen tea was a substitute for black tea (*Camellia sinensis*). The berries were traditionally soaked in brandy, and the resulting extract was taken to stimulate appetite, as a substitute for bitters. All false-wintergreens contain methyl salicylate, a close relative of aspirin that has been used to relieve aches and pains. These plants have been widely used in the treatment of painful inflamed joints resulting from rheumatism and arthritis (see Warning).

Studies suggest that oil of wintergreen is an effective painkiller, and it has numerous commercial applications: it provides fragrance to various products such as toothpastes, chewing gum and candy and is used to mask the odours of some organophosphate pesticides. It is a flavouring agent (at no more than 0.04 percent) and an ingredient in deep-heating muscle creams.

The oil is also a source of triboluminescence, a phenomenon in which a substance produces light when rubbed, scratched or crushed. The oil, mixed with sugar and dried, builds up an electrical charge that releases sparks when ground, producing the Wint-O-Green Lifesavers phenomenon. To observe this, look in a mirror in a dark room and chew the candy with your mouth open!

EDIBILITY: edible with caution (see Warning)

FRUIT: Mealy to pulpy, fleshy, berry-like capsules with a mild, wintergreen flavour.

SEASON: Flowers May to June. Fruits ripen late August into September.

DESCRIPTION: Delicate, creeping evergreen shrublets. Leaves leathery, small. Flowers white to greenish or pinkish.

Hairy false-wintergreen (*G. hispidula*) has tiny, stiff, flat-lying, brown hairs on its stems and lower leaf surfaces. Leaves very small (3–10 mm long). Flowers tiny (2 mm wide), 4-lobed, bell shaped, born singly in leaf axils. Berries white, small (generally less than 5 mm in diameter), on a very short stalk, persisting through fall. Grows in cold, wet bogs and coniferous forests in montane and subalpine zones across the prairie provinces. Associated with acidic soils and often grows in mosses under conifers or on rotting logs, along the edges of swamps or bogs. Also called: **creeping snowberry.**

Alpine false-wintergreen (*G. humifusa*)

False-wintergreens

Alpine false-wintergreen (*G. humifusa*) has 1–2 cm long, glossy leaves, rounded to blunt at tips, with pinkish- or greenish-white, 5-lobed, 3–4 mm wide flowers and scarlet, pulpy, 5–6 mm wide berries. Berries are drier and not as palatable as other *Gaultheria* species. Grows in moist to

Wintergreen (*G. procumbens*)

Wintergreen (*G. procumbens*)

Hairy false-wintergreen (*G. hispidula*)

TREES & SHRUBS

wet, subalpine to alpine meadows in AB. Also called: alpine wintergreen, creeping wintergreen.

Wintergreen (*G. procumbens*) is 10–20 cm tall, with thick, shiny, oval leaves, 2.5–5 cm long, and small, oval 5-lobed flowers from April to May. Dry, red berries dangle beneath the leaves. Grows in poor or sandy soils under evergreens or in oak woods from southeastern MB to the east.

> **WARNING:** *Oil-of-wintergreen (most concentrated in the berries and young leaves of wintergreens) is considered highly toxic if taken in large quantities, especially for children. It should never be taken internally, except in very small amounts. Avoid applying the oil when you are hot, because dangerous amounts could be absorbed through the open pores of your skin. It is known to cause skin reactions and severe (anaphylactic) allergic reactions. People allergic to aspirin should not use false-wintergreen or its relatives.*

Alpine false-wintergreen (*G. humifusa*)

Cranberries   *Vaccinium* spp.

Lingonberry (*V. vitis-idaea*)

Like many other species of wild berries with domesticated counterparts, wild cranberries are small but packed with a flavour that seems disproportionate to their size. The tartness of cranberries gives them an enviable versatility for sweet, sour and savoury dishes. Who could imagine Thanksgiving dinner or many juices, desserts or baking without them? These berries, which also have a long history of medicinal use in treating kidney and urinary ailments, are touted today for their strong antioxidant properties. (Note: high bush cranberries, while tart and cranberry-tasting, are in the honeysuckle family and are therefore treated in a separate account.)

Cranberries can be very tart, but they make a refreshing trail nibble. They also make delicious jams and jellies, and they can be crushed or chopped to make tea, juice or wine. A refreshing drink is easily made by simmering berries (crush them first to allow the juice out more easily) with sugar and water or, more traditionally, by mixing cranberries with maple sugar and cider. Cranberry sauce is still a favourite with meat or fowl, and the berries add a pleasant zing to fruit salads, pies and mixed fruit cobblers. Cranberries are also a delicious addition to pancakes, muffins, breads, cakes and puddings.

Firm, washed berries keep for several months when stored in a cool place. They can also be frozen, dried or canned.

First Nations sometimes dried berries for use in pemmican, soups, sauces and stews. Freezing makes cranberries sweeter, so they were traditionally collected after the first heavy frost. Because they remain on the shrubs all year, cranberries can be a valuable survival food rich in vitamin C and antioxidants. These low-growing berries are difficult to collect, so some people combed them from their branches with a salmon backbone or wooden comb.

Cranberry juice has long been used to treat bladder infections. Research shows that these berries contain arbutin, which prevents some bacteria from adhering to the walls of the bladder and urinary tract and causing an infection. Cranberry juice also increases the acidity of the urine, thereby inhibiting bacterial activity, which can relieve infections. Commercial cranberry juice or cocktail blends are not appropriate for this treatment, however, as the juice is highly processed, often diluted with other juices and is highly sweetened (sugars will feed the problem bacteria and can make the existing condition worse). Increased acidity can also lessen the urinary odour of people suffering from incontinence. The tannins in cranberry have anticlotting properties and are able to reduce the amount of dental plaque-causing bacteria in the mouth, thus are helpful against gingivitis.

Research has shown that cranberries contain antioxidant polyphenols that may be beneficial in maintaining cardiovascular health and immune function, and in preventing tumour formation. Although some of these compounds have proved extraordinarily powerful in killing certain types of human cancer cells in the laboratory, their effectiveness when ingested is unknown. There is also evidence that cranberry juice may be effective against the formation of kidney stones.

Cranberries were traditionally prescribed to relieve nausea, fevers, and sore throats, to ease cramps in childbirth and to quiet hysteria and convulsions. Crushed cranberries were used as poultices on wounds, including poison-arrow wounds. The Inupiat Inuit used a wrapped cloth containing mashed berries to treat sore throats and crushed berries to treat itchy skin conditions such as chicken pox and measles. The mashed fruit, mixed with oil, was fed to convalescents to help them recover strength. The red pulp (left after the berries have been crushed to make juice) can be used to make a red dye. Lingonberry is popular in Sweden as a digestive aid, in jams, jellies, pies and other baking, juice, wine liqueur, herbal tea, and as "cranberry sauce."

Bog cranberry (*V. oxycoccos*)

Cranberries

EDIBILITY: highly edible

FRUIT: Berries bright red (sometimes purplish) when ripe, sour, 6–10 mm wide. Tart and delicious, said to be best after the first frost, and can be foraged when snows melt in the spring—if the wildlife have left any!

SEASON: Flowers June to July. Fruits ripen late July to September and may persist on plants throughout winter.

DESCRIPTION: Dwarf, low-spreading deciduous shrub, mostly less than 20 cm tall, often trailing, with small, nodding, pinkish flowers.

Bog cranberry (*V. oxycoccos*) has slender, creeping stems, with small (mostly less than 1 cm long), pointed, glossy leaves. Flowers are distinctive, 4 petals separated almost to the base, the petals curved strongly backward (like little shooting stars), flowers appearing terminal on stems. Fruit a deep red, globose cranberry, juicy, about 5–12 mm wide. Bog cranberry grows across the prairie provinces and predictably inhabits boggy areas. Also called: small cranberry • *Oxycoccus oxycoccos, O. quadripetalus, O. microcarpus*.

Grouseberry (*V. scoparium*) is a low, creeping, broom-like shrub to 25 cm tall, with many slender, squarish, pale green, strongly angled branches. Leaves thin, finely toothed, deciduous, alternate, ovate, 6–12 mm long. Flowers tiny, 3–4 mm long, 4 petals fused into urn-shaped blooms, pink, singly from leaf axils. Berries single, small (3–5 mm wide), bright red. Grows on foothill, montane and subalpine slopes in southwestern AB.

Lingonberry (*V. vitis-idaea*)

Grouseberry (*V. scoparium*)

Grouseberry (*V. scoparium*)

TREES & SHRUBS

Bog cranberry (*V. oxycoccos*)

Cranberry Chicken

Serves 5

3 lbs chicken • ¼ cup flour • ½ tsp salt
¼ cup cooking oil • 1½ cups fresh cranberries
½ cup sugar • 1 Tbsp grated orange zest
½ cup fresh orange juice • ¼ tsp ground ginger

Cut chicken into serving-sized pieces, and coat with flour and salt. Heat oil in a cast-iron skillet. Add chicken pieces and brown on both sides, being careful to cook the chicken fully. Combine remaining ingredients in a saucepan, bring to a boil and pour it over the chicken in the skillet. Cover skillet, reduce heat and simmer 30 to 40 minutes until chicken is tender.

Lingonberry (*V. vitis-idaea*) is a low-spreading shrub, up to 15–30 cm in length, with rounded or slightly angled branches. Leaves evergreen, 6–15 mm long, blunt, leathery, with dark dots (hairs) on a pale lower surface. Flowers 4-petalled, fused into small urn-shaped nodding blooms, pinkish in colour, 1 to several in terminal clusters. Fruit bright red cranberries, 6–10 mm wide. Grows in all prairie provinces, preferring acidic soils in sunny mountain meadows, peat moors, dry woods, foothill, montane, subalpine and alpine slopes. Also called: mountain cranberry, low bush cranberry, rock cranberry, cowberry, partridge berry.

Bog cranberry (*V. oxycoccos*)

Blueberries *Vaccinium* spp.

Dwarf blueberry (*V. caespitosum*)

Blueberry fruit tend to be blue in colour, hence the common name of this group of plants. The fruit of this species is generally sweet rather than tart/sour (cranberries) or sweet/tangy (huckleberries). Wild blueberries are simply delicious! Rich in vitamin C and natural antioxidants, these fruit are both beautiful to look at and good for you to eat. Among a myriad of other uses, the sweet, juicy berries can be consumed fresh from the bush, added to fruit salads, and cooked in pies, tarts and cobblers, made into jams, syrups and jellies, or crushed and used to make juice, wine, tea and cold drinks. Blueberries also make a prime addition to pancakes, muffins, cakes and puddings as well as to savoury treats like chutneys and marinades.

These wild fruits were widely used by First Nations across Canada, either fresh, dried singly or mashed and formed into cakes alone or with other fruits. To make dried cakes, the berries were cooked to a mush to release the juice, spread into slabs and dried on a rack (made from wood, rocks or plant materials) in the sun or near a fire. Often, the leftover juice was slowly poured onto the drying cakes to increase their flavour and sweetness.

Because blueberries grow close to the ground, they can be difficult and time-consuming to collect, so some people developed a method of combing them from the branches with a salmon backbone, wooden comb or rake. While this method is efficient, it results in baskets full of both berries

TREES & SHRUBS

and small hard-to-pick-out blueberry leaves. The savvy solution developed for this problem was to place a wooden board at a medium angle and slowly pour the berry/leaf mix from the top of the board: the berries (which are round) roll down the board to a basket waiting below, but the leaves (which are flat) stick to the board and stay put. After 2 or 3 rollings, the picker ends up with a basket of pure berries at significantly less effort than would have been required to pick the leaves out individually. The only drawback to this method is that the occasional green berry also gets picked, but these are easily removed by hand.

While most people only associate blueberries with a delicious fruit, there are many historical medicinal uses for other parts of this wide-ranging plant. Blueberry roots were boiled to make medicinal teas that were taken to relieve diarrhea, gargled to soothe sore mouths and throats, or applied to slow-healing sores. Bruised roots and berries were steeped in gin, which was to be taken freely (as much as the stomach and head could tolerate!) to stimulate urination and relieve kidney stones and water retention. Blueberry leaf tea and dried blueberries have historically been used like cranberries to treat diarrhea and urinary tract infections. Blueberries contain anthocyanins, which may reduce leakage in small blood vessels (capillaries), and they have been suggested as a safe and effective treatment for water retention during pregnancy, for hemorrhoids, varicose veins and similar problems. They have also been recommended to reduce inflammation from acne and other skin problems and to prevent cataracts. Blueberry leaf tea has been used by people suffering from hypoglycemia and by some diabetics to stabilize and reduce blood sugar levels and to reduce the need for

Dwarf blueberry (*V. caespitosum*)

Oval-leaved blueberry (*V. ovalifolium*)

Blueberries

insulin. The leaves of blueberry were sometimes dried and smoked, and the berries have been used to dye clothing or porcupine quills a navy blue colour.

EDIBILITY: highly edible

FRUIT: Berries round, 5–8 mm wide, bluish coloured, growing in clusters, usually with a greyish bloom. Most blueberries have a small "crown" at the bottom end of the berry, which is a leftover from the flower that was pollinated to form the fruit.

SEASON: Flowers May to July. Fruits ripen July to September.

DESCRIPTION: Low, often matted shrub with thin, elliptical leaves 1–3 cm long. Flowers whitish to pink, nodding, urn-shaped, 4–6 mm long.

Lowbush blueberry (*V. angustifolium*)

Lowbush blueberry (*V. angustifolium*)

TREES & SHRUBS

Lowbush blueberry (*V. angustifolium*) is a low shrub, to 60 cm tall, but usually less than 35 cm, with toothed leaves. White, bell-shaped flowers 5 mm long, from May to June. Small, dark blue or black berries (not with a bloom), 1.2 cm across, in clusters at the tips of twigs. Grows in abundance in dry, open barrens, peats and rock in MB. This province is the main source of wild harvested blueberries in eastern Canada.

Dwarf blueberry (*V. caespitosum*) is a low, usually matted shrub, 10–30 cm tall, with rounded, yellowish to reddish branches and shiny, light green leaves with distinct teeth on upper half. Its 5-lobed flowers produce 1–4 berries in leaf axils, from August to September. Grows at all elevations in dry to wet forests, bogs, meadows, rocky ridges and tundra across the prairie provinces.

Velvet-leaved blueberry (*V. myrtilloides*) is a low shrub, to 50 cm tall, with densely hairy (velvety) branches, especially when young. Leaves 2–4 cm long, soft-hairy with smooth edges. Flowers in clusters at branch tips, 3–5 mm long. Dark bluish black to dark red fruits, from August to October. Grows at montane elevations in dry to moist forests and openings and bogs across the prairie provinces.

Oval-leaved blueberry (*V. ovalifolium*) is a tall shrub, to 2 m tall, with hairless, angled branches. Leaves are entire or only sparsely toothed. Purple berries with a whitish bloom, from early July to September. Grows at low to subalpine elevations in dry to moist forests, openings and bogs in AB.

> **WARNING:** *Blueberry leaves contain moderately high concentrations of tannins, so they should not be used continually for extended periods of time.*

Velvet-leaved blueberry (*V. myrtilloides*)

Velvet-leaved blueberry (*V. myrtilloides*)

Blueberries

Bog blueberry (*V. uliginosum*) is a low, spreading, deciduous, perennial shrub, 10–60 cm tall (but can be as short as 2.5 cm in areas of heavy snow where the shrub is crushed flat each winter). Branches are rounded, brownish. Leaves 3 cm long, fuzzy to smooth, elliptical to oval, narrow and broadest toward the tip, dull/whitish green, toothless. Flowers solitary or paired, 4 lobed, white or pinkish, urn-shaped, up to 6 mm long. Berries dark blue to blackish with a whitish bloom, to 9 mm in size. Inhabits low-elevation bogs, boggy forests, subalpine heath and alpine slopes/tundra across the prairie provinces. Also called: bog bilberry, western huckleberry • *V. occidentale*.

Vaccinium **spp.**
Note that common names can be confusing, especially with the large variety of Vaccinium *species found across Canada. Blueberries, cranberries and huckleberries are all closely related plants in this plant family. In North America there are approximately 35 different* Vaccinium *species, but hybridization is common in the genus so the true number of varieties is probably greater. As a general rule, species of* Vaccinium *with blue fruits are called blueberries, and taller shrubs with fruits that aren't blue are called huckleberries. Shorter* Vaccinium *species with red berries and a distinctive tart flavour are commonly referred to as cranberries. However, as illustrated by black huckleberry, common names do not necessarily follow this botanical protocol. Another example is high bush cranberry,* Viburnum edule, *which in the honeysuckle family and is not a "true" cranberry at all, despite its red colour and tart flavour. Rest assured, though, none of these species are poisonous and they are all delicious to eat!*

Bog blueberry (*V. uliginosum*)

Oval-leaved blueberry (*V. ovalifolium*)

TREES & SHRUBS

Bog blueberry (*V. uliginosum*)

Blueberry Cobbler

1 cup flour • 2 Tbsp sugar • 1½ tsp baking powder
¼ tsp salt • 1 tsp grated lemon zest • ¼ cup butter
1 beaten egg • ¼ cup milk • 2 Tbsp cornstarch
½ cup sugar • 4 cups fresh blueberries (or huckleberries)

Preheat oven to 425° F. Sift together flour, sugar, baking powder and salt. In another bowl, mix zest, butter, egg and milk, then pour slowly into the dry mix, stirring until just moistened.

Mix cornstarch and sugar together, and toss with the fruit. Pour this mixture into the bottom of an 8 x 10-inch glass or ceramic baking dish (avoid metal dishes because the acid in the fruit might turn it rusty and impart a nasty flavour to the cobbler). Drop the topping in spoonfuls on top of the fruit, covering the surface as evenly as possible (some exposed areas of fruit are fine). Bake, uncovered, for 25 minutes or until light brown.

Fruit Popsicles

Makes 8 to 12 popsicles

An easy and popular treat for adults and kids at any time of the year!

4 cups wild berries • 1 cup plain yogurt or light cream
1 cup white sugar • 1 cup orange or other fruit juice (optional)

Blend all ingredients together, pour into the compartments of a popsicle maker and place in freezer until frozen.

Huckleberries *Vaccinium* spp.

Black huckleberry (*V. membranaceum*)

Huckleberries are delicious and well worth identifying and eating. These berries are generally sweet but can be tart and strong flavoured (blueberries, on the other hand, tend to be purely sweet and cranberries sharply tart). Despite huckleberries and blueberries being separated by two distinct sets of common names, botanists do not make a formal distinction between these two fruits. Huckleberry fruit are usually blackish and glossy while blueberries are generally blue in colour with a whitish bloom on the fruit.

Huckleberries can be used like domestic blueberries, eaten fresh from the bush, added to fruit salads, cooked in pies and cobblers, made into jams and jellies or crushed and used in cold drinks. They are also delicious in pancakes, muffins, cakes and puddings. You can use huckleberries wherever cranberries or blueberries are called for in a recipe, both sweet and savoury. Black huckleberries are collected in large quantities even today in open, subalpine sites (such as old burns), and in some regions they are sold commercially. Dried huckleberry leaves and berries also make excellent teas, and huckleberries are relished by bears and other wildlife.

Huckleberries were considered good for the liver by some First Nations and were eaten as a ceremonial food to ensure health and prosperity for the coming season. These fruit were eaten fresh or dried for winter use and trade.

The small size of many huckleberries makes picking them rather time-consuming. Try using a (clean) comb to rake the berries into a basket, hat or other container to speed up this process. They can grow quite abundantly in alpine areas, so a little persistence can yield a good haul.

EDIBILITY: highly edible

FRUIT: Berries range in colour from red to purple to black.

SEASON: Flowers April to June. Berries ripen July to September.

DESCRIPTION: Deciduous or evergreen shrubs, 40 cm–2 m tall, with alternate, 2–5 cm long, finely saw-toothed leaves. Flowers various shades of pink, round to urn-shaped, 4–6 mm long, nodding on single, slender stalks. Berries 6–10 mm across.

Black huckleberry (*V. membranaceum*) is a deciduous shrub with purplish black berries. Branches greenish when young but turn greyish brown with age, at most slightly angled. Grows in dry to moist forests at montane to subalpine elevations on open or wooded slopes in AB. Also called: thinleaf huckleberry, black mountain huckleberry.

LITERARY REFERENCE: *The small, dark, rather insignificant fruit of this species became a metaphor in the early part of the 19th century for something humble or minor. This later came to mean somebody inconsequential, an idea that was the basis for Mark Twain's famous character, Huckleberry Finn. In an 1895 interview, Twain revealed that he wanted to establish the boy as "someone of lower extraction or degree, than Tom Sawyer." The modern word "huckleberry" is an American distortion of the British "whortleberry," a name that originated from the Anglo Saxon words* wyrtil, *meaning "little shrub," and* beri, *meaning "berry."*

Black huckleberry (*V. membranaceum*)

Black huckleberry (*V. membranaceum*)

Huckleberries

False huckleberry (*Menziesia ferruginea*) has small urn-shaped flowers that look very similar to *Vaccinium* species and small pink berry-like "fruit" on the underside of its leaves. However, this "berry" is actually the fruiting body of a fungus (*Exobasidium vaccinii*). While all parts of this plant are poisonous, these fungal "berries" are apparently edible. Grows in AB.

WARNING: *The plant false huckleberry (*Menziesia ferruginea*) is poisonous, as are its fruit. The edible "berry" sometimes found on the underside of its leaves is the fruiting part of a parasitic fungus.*

False huckleberry (*Menziesia ferruginea*)

False huckleberry (*Menziesia ferruginea*)

Huckleberry Relish

Makes about 2 cups

1 cup water
2 cups huckleberries
½ cup sugar
1 onion, thinly sliced
½ cup apple cider or white wine vinegar
4 cloves garlic, finely chopped

Put water, berries and sugar in a heavy-bottomed saucepan and bring the mixture slowly to a boil. Cook and stir for about 5 minutes until the fruit softens. Remove pan from heat and stir in onion, vinegar and garlic. Mix well and allow to cool on the stove with a lid on. When cooled, spoon into a jar, cover with a lid and place in the refrigerator. Keeps well for 2 weeks.

TREES & SHRUBS

Tom's Huckleberry Pie

Makes 1 double-crust pie

The secret to a good, crispy pastry that is not tough and "dough-like" is to keep all your ingredients cool. Warm ingredients melt the small fat globules, causing them to mix too completely with the flour and resulting in chewy pastry. Leftover pastry trimmings make excellent little tartlets if rolled out again and put into the bottom of muffin tins, and filled with any extra huckleberry filling or jam out of the fridge.

Pastry
2 cups all-purpose flour • ½ tsp salt • ⅔ cup vegetable shortening • ⅓ cup cold milk

Filling
3 cups black huckleberries • ¼ cup water or freshly squeezed orange juice
1 cup sugar • 3 Tbsp cornstarch

For the pastry, sift the flour and salt together, then use 2 butter knives or a pastry cutter to cut the shortening into the flour mixture until the butter pieces are the size of small peas. *Avoid using your hands at this stage because their warmth will cause the butter to melt.* Gradually stir in the cold milk, then quickly shape the dough into 2 balls. Wrap them in plastic film, press into flat rounds and refrigerate immediately.

For the pie filling, mash the huckleberries with the water and put into a medium saucepan. Mix the cornstarch and sugar together and stir well into the cold berries and water. *Do not heat the berries and water first because the cornstarch will cook prematurely and go all nasty and lumpy!* Bring mixture slowly to a boil, stirring constantly to avoid the cornstarch sticking or getting lumps. Simmer until noticeably thick, about 4 to 5 minutes, then take the saucepan off the burner.

Preheat oven to 350° F. Take the pastry out of the fridge, spread a thin layer of flour on a work surface, and roll the pastry until it is approximately ¼ inch thick. Place it into an 8-inch pie tin, cutting any extra off from around the edges. Then roll out the second half of your dough into a similar-sized round. Fill the pastry shell with the huckleberry mixture, then carefully place the second round on top. Gently push the edges of the top and bottom pastry crusts together (you may need to lightly wet one edge to get it to stay together), and prick a few holes with a fork in the top to allow steam to escape during cooking. Bake for approximately 50 minutes.

False huckleberry (*Menziesia ferruginea*)

Whortleberry *Vaccinium myrtillus*

Also called: low bilberry, bilberry, common bilberry • *V. oreophilum*

Whortleberry (*V. myrtillus*)

This species is a close relative of the blueberry but (like huckleberries) differs in a number of ways. Whortleberries produce single or paired fruit rather than clusters. The fruit is smaller than that of the blueberry and generally has a stronger taste. It is also darker in colour, usually appearing near black with a slight shade of purple on the outside. The fruit is red or purple inside (the interior of blueberry fruit is light green). One of the reasons that whortleberries are not often sold commercially is that the fruits are softer and juicier than blueberries and have a thinner skin, making them difficult to transport. The thin skin, deep colour and juiciness mean that it is easy to stain your fingers (and lips!) when harvesting the raw fruit. Whortleberries are sweet but have a slightly acid edge, and some reports suggest that they are best eaten cooked rather than raw. These tasty berries can be used in many ways, including pies, jams, tarts, syrups and jellies. The berries also have a long history of use in wine-making and for juice, cordial, liqueur and sorbet.

As a dark purple to black fruit, whortleberries contain high levels of anthocyanin pigments (these pigments cause the blue, red and violet colours in fruits and flowers). Anthocyanin is important medicinally and has been linked experimentally to lowered risk for several diseases, for example those of the heart and cardiovascular systems, ocular disorders, diabetes and some cancers. Supplements containing whortleberry extracts (mainly of the berries and leaves) can commonly be found in health food stores and are recommended for circulatory problems, diarrhea and diabetes among other complaints. Whortleberry leaves contain glucoquinine, a compound that has been shown to have a mild effect in lowering blood sugar levels. The leaves, which are generally prescribed as a tea, also have astringent, tonic, anti-inflammatory and antiseptic qualities. However, whortleberry leaves contain relatively high levels of tannins, and animal studies suggest that they can cause anemia, disturb the gastrointestinal system and adversely effect absorption of nutrients, so they are not recommended for use.

EDIBILITY: highly edible

FRUIT: Globe-shaped, dark red to bluish black, 4–9 mm.

SEASON: Flowers May to July. Fruits ripen July through September.

DESCRIPTION: Small, many-branched shrub growing to 30 cm high and up to 30 cm wide, forming open colonies. Stems strongly angled, greenish brown, minutely covered in fine hairs or down along grooves. Leaves broadly 1–3 cm long, alternate, elliptic to ovate, bright green, sharply serrated margins. Flowers greenish white, cream or pink, glabrous, 2–3 mm, globose-shaped, in leaf axils, solitary, on nodding stalks. Inhabits damp, nutrient-poor, acidic, sandy or loamy soils in open woods or coniferous forests, heaths, morraines and other disturbed soils, moors and bogs in AB. Note that the common name "whortleberry" also refers to *Arctostaphylos alpina*.

Riverbank Grape *Vitis riparia*
Also called: frost grape

Riverbank grape (*V. riparia*)

The fruit of riverbank grape (from which the concord grape is derived) is small and tart, but juicy and flavourful. Many First Nations valued the berries (eaten fresh or preserved for winter) as a food source. The grapes are said to be best when harvested after the first frost, which makes them taste sweeter, but they are often consumed rapidly by birds, so gatherers need to be quick! In more recent times, the berries have been made into jelly and wine. The wine of our native grapes reportedly has a musky, "foxy" flavour. Wine and grapeseed extract contain the compound resveratrol, which has beneficial cardiovascular and antidiabetic properties.

The leaves can be pickled to use in making dolmades, a Greek delicacy of rice and meat wrapped in vine leaves. Did your grandmother insist on stuffing a grape leaf in each jar of pickles she made? The leaves contain varying amounts of a natural inhibitor that reduces the effects of a softening enzyme present on mouldy cucumber blossoms. Adding a grape leaf to each jar of homemade pickles is a traditional practice that results in pickles that are less likely to go soft.

The winter vines, when twisted together, make a solid and decorative base for Christmas wreaths. Riverbank grape is a key parent species in breeding modern grape varieties for fruit and wine that are disease resistant and cold tolerant.

EDIBILITY: highly edible

FRUIT: Tight cluster of small (10–12 mm across), black, spherical berries with a waxy coating, giving them a bluish cast, contains 2–6 small, oval seeds each.

SEASON: Flowers May to June. Fruits ripen August to September.

DESCRIPTION: A woody, deciduous vine either climbing by means of tendrils or trailing on the ground. Leaves alternate, 7–20 cm long, toothed, mostly 3-lobed (some unlobed), hairless or slightly hairy below, coarsely toothed. Flowers greenish, inconspicuous; compact in pyramidal clusters. Inhabits moist thickets in southern SK and MB.

Riverbank Grape Jelly

Makes about 4 x 1 cup jars

4 cups ripe riverbank grapes, de-stemmed
1 cup water • unsweetened apple juice
1 x 2 oz package pectin • 5 cups sugar

Clean and crush grapes. Place water and crushed grapes into a heavy-bottomed saucepan and cook slowly for about 20 minutes until fruit is softened and letting its juices go. Strain mixture through a jellybag or cheesecloth. *Do not squeeze the cloth or it will cause sediments to run into the juice, resulting in a cloudy jelly.* If the resulting liquid does not measure 4 cups, add some apple juice until 4 cups liquid is reached. Add liquid back to clean saucepan. Mix in pectin and bring the mixture to a full boil. Add sugar and continue stirring, scraping the bottom of the pot so that the jelly doesn't burn. Boil hard for 3 minutes, then remove from heat and pour into hot sterilized jars. Wipe jar edges clean and seal with hot sterilized lids (waterbathing them for 5 minutes will also seal jars).

Sarsaparillas *Aralia* spp.

Wild sarsaparilla (*Aralia nudicaulis*)

These fragrant plants are related to ginseng and have a warm, aromatic, sweetish taste that is most intense in the rhizomes and berries. The berries have traditionally been used to flavour beer and to make wine (similar to elderberry wine), and a tea was sometimes made from the seeds. However, descriptions of the edibility of the berries range from not palatable to poisonous. The fruit of this species have also been used to make jelly, but this is not recommended (see Warning).

The berries and rhizomes have a long history of being boiled to make medicinal teas and syrups or soaked in alcohol to make tinctures. These medicines were used to treat many different complaints, ranging from stomachaches to rheumatism and syphilis. Wild sarsaparilla was widely used in patent medicines in the late 1800s. Spikenard is a particularly attractive member of this species, with large decorative leaves that turn golden in fall and a striking flower spike that can reach 2 m high. It makes a nice addition to the moist, shade garden.

EDIBILITY: not palatable, edible with caution (toxic)

Hairy sarsaparilla (*A. hispida*)

FLOWERING PLANTS

FRUIT: Dark red to purple berries growing in clusters on a terminal stalk.

SEASON: Flowers May to August. Fruits ripen August to September.

DESCRIPTION: Perennial shrubs or herbs growing from rhizomes. Leaves large and compound. Flowers whitish, growing in terminal clusters on stems.

Hairy sarsaparilla (*A. hispida*) grows to 1 m tall, with sharp, stiff bristles at the base and covering the stems. Leaves twice compound with oval, toothed leaflets. Flowers small, greenish white, grow in globe-shaped clusters from June to August. Fruits are dark, foul-smelling berries. Grows in sandy, open woods from SK to MB.

Wild sarsaparilla (*A. nudicaulis*) grows to 70 cm tall, with a long, horizontal rhizome. Leaf blades horizontal, with 3 major divisions, each of these divided into 3–5 oval leaflets, 2–3 cm long, finely toothed. Flowers greenish white, 2–3 mm long, forming 2–7 (usually 3) round, 2–5 cm-wide clusters, from May to June, usually hidden under the leaf. Fruits dark purple berries, 6–8 mm wide. Grows in moist, shaded forests throughout the prairies.

Spikenard (*A. racemosa*) grows to 2 m tall, with dark green or reddish stems. Leaves 3 times compound with 6–21 toothed, heart-shaped leaflets. Flowers small, whitish, growing in small clusters along a branching raceme, from June to August. Grows in rich, moist woods from southern MB eastward.

Wild sarsaparilla (*A. nudicaulis*)

Spikenard (*A. racemosa*)

WARNING: *Some people have reported being very sick after eating wild sarsaparilla berries, and consuming this fruit is not recommended.*

Common Barberry *Berberis vulgaris*

Common barberry (*B. vulgaris*)

This dense, viciously thorny shrub provides edible and medicinal benefits that have a long history of use. Barberry fruits are acceptable raw in small quantities, but they are very acidic and instead make rather excellent preserves or jellies. In Iran, dried *Berberis* berries, called zereshk, are widely used and impart a tart flavour to chicken and rice dishes. A popular candy in the Ukraine of the same name is also made from the berries. A rich source of vitamin C and the soluble fibre pectin, barberry berries make a refreshing lemonade-like cold beverage when sweetened.

Barberry contains a number of well-studied alkaloids, especially berberine (found mostly in the roots), which is antimicrobial, anti-inflammatory and astringent. It has been tested for its potential usefulness in treating diabetes, prostate cancer, cardiac arrhythmia and leukemia, although not enough research has been conducted in humans. By itself, berberine is a relatively weak antibiotic. However, an extract that contains barberry's other components, such as the isoquinoline alkaloids berbamine and oxyacanthine, displays significantly stronger antibacterial activity, as well as effects against amoebas and trypanosomes. These alkaloids are poorly absorbed through the digestive tract, so they are particularly useful against enteric infections, such as bacterial dysentery, and parasite infections. In this regard, herbalists do not recommend that wild licorice be consumed at the same time because it is surmised that it nullifies the effects of barberry.

The clinical use of purified berberine today is mainly to counteract bacterial diarrhea, and in eye drops to treat ocular trachoma infection, hypersensitive eyes, inflamed eyelids and conjunctivitis. Berberine can be poisonous if taken in large doses (see Warning).

A mild laxative can be made from an infusion of the berries mixed with wine. A decoction of the berries or root bark makes an effective mouthwash or gargle for mouth and throat complaints. Fresh berry juice was thought to strengthen the gums and relieve pyorrhea when brushed directly on them. The shrubs are sometimes planted as hedges and as barriers under vulnerable windows to deter trespassers. They tolerate trimming well. Barberry is the intermediate host for wheat rust and was the focus of an extensive eradication effort that lasted from 1918 to 1975.

EDIBILITY: edible

FRUIT: Small berries, elliptical and scarlet when mature, usually long and narrow, resembling a bar (hence the common name barberry).

SEASON: Flowers May to June. Fruits ripen August to September.

DESCRIPTION: Perennial, deciduous shrub growing 90 cm–3 m tall. Branches grey. Long shoots erect, branched or unbranched, with simple or 3-spined thorns 3–30 mm long. Short shoots in thorn axils produce leaves to 10 cm long, margins entire or spiny. Leaves 2.5–7.5 cm long and bristle-toothed. Flowers pendulous clusters of 10–20 per head, yellow, 3–6 mm long, with 6 petals and sepals. This introduced Eurasian species is widely naturalized in disturbed spaces (pastures, roadside thickets) in southern MB.

Prickly-pear Cacti *Opuntia* spp.

Brittle prickly-pear cactus (*O. fragilis*)

Prickly-pear cacti were widely used for food in western Canada, though the fruits of Canadian cacti are smaller and less fleshy than those of their southern relatives. The flavour ranges from bland to sweet to sour and has been likened (at best) to sweet pomegranates or cucumber. The spines on the fruits were removed by peeling, sweeping piles of fruit with big sagebrush branches, singeing or simply by picking them off (with fingers protected by deerskin tips). The fruits were then split to remove the seeds and eaten raw (alone or with other fruits), cooked into stews and soups as a thickener or dried and ground for later use. More recently, the sweet flesh has been added to fruit cakes or canned with other fruits or juice. Berries can also be boiled whole and strained to make pretty-coloured jellies or syrups. Raw cactus stems were often eaten when there was a shortage of food. Young segments were boiled and peeled to remove their spines, and the pulpy flesh was fried. Alternately, roasted or pit-cooked stems were squeezed until the edible inner part popped out. In more modern times, cactus stems have also been pickled or candied.

The peeled, mucilaginous stems were used for dressing wounds or were mashed and placed on aching backs. Stems were also boiled to make a medicinal drink for relieving diarrhea and lung problems and for treating people who could not urinate. Split stems were placed in containers of muddy water, where they exuded large amounts of mucilage, which cleared the water and made it drinkable. Freshly peeled stems were sometimes rubbed over painted hides to fix the colours. The fruit provided a pink- to red-coloured dye. The gum from the stem was sometimes used as an adhesive. When forage was limited, the spines were singed off and cactus stems were fed to livestock.

More recently, studies have suggested that the juice may be effective in lowering blood sugar levels in diabetics, especially those with chronic hyperglycemia.

WARNING: *It goes without saying that you should always protect your hands with good gloves when collecting these spiny plants!*

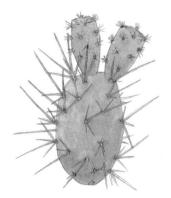

Brittle prickly-pear cactus (*Opuntia fragilis*)

EDIBILITY: edible

FRUIT: Fleshy (though often rather dry), seedy, somewhat spiny, oval berries 1.5–2.5 cm long.

SEASON: Flowers June to July. Fruits ripen August to September.

DESCRIPTION: Spiny, perennial herbs from fibrous roots, 5–40 cm tall, with thick, fleshy, segmented stems. Leaves reduced to spirally arranged, starburst clusters of short bristles and rigid, barbed, 1–5 cm-long spines. Flowers yellow (sometimes pinkish when older), broadly bell-shaped, with many paper-thin, overlapping petals, solitary at branch tips.

Brittle prickly-pear cactus (*O. fragilis*) has strongly barbed spines in white-woolly tufts on small (less than 5 cm long), rounded stem segments that detach easily. Grows in dry, open ground in prairie to foothills across the southern prairie provinces.

Plains prickly-pear cactus (*O. polyacantha*) has slightly barbed spines in brown-woolly or hairless tufts on large (often 5–12 cm long), flattened stem segments that do not break apart easily. Grows in dry, open prairie and on sandhills and rocks from AB to SK.

Plains prickly-pear cactus (*Opuntia polyacantha*)

Strawberry Blite *Chenopodium capitatum*

Also called: strawberry spinach, blite goosefoot

Strawberry blite (*C. capitatum*)

This genus includes several plants of minor to moderate importance as food crops, including quinoa (*C. quinoa*) and kañiwa (*C. pallidicaule*). The fruits of strawberry blite are edible and nutritious but rather bland; some indigenous groups considered them inedible. They can be added to salads, eaten raw or cooked. The leaves are satisfactory eaten raw, although not too much should be consumed because of possible toxicity (see Warning). They are best cooked, boiled or steamed for 10–15 minutes and served like spinach. The leaves are not as tasty as spinach and lose quite an amount of bulk upon cooking, but they are excellent if stronger flavoured greens are added. If eaten with beans, they reportedly act as a carminative and prevent "wind" and bloating. The leaves are quite nutritious, containing (per 100 g dry weight) 260 calories, 24 g protein, 5 g fat, 45 g carbohydrates, 15 g fibre, plus significant amounts of calcium, phosphorus, iron and vitamin A.

Strawberry blite was traditionally used in lotions for treating black eyes and head bruises. Lung congestion was treated with the juice expressed from the seeds and an infusion of the entire plant. A red-coloured dye can be obtained from the fruits, which some tribes used as a face paint and for decorating porcupine quills, basket materials and hides.

EDIBILITY: edible

FRUIT: Bright red, and technically a tiny achene, containing a single seed, resembling a strawberry, clustered into balls to 12 mm across, from leaf axils or in spikes along stem tips.

SEASON: Flowers July to August. Fruits ripen August to September.

DESCRIPTION: Succulent, erect, annual herb from a taproot, 20 cm–50 cm tall. Stem often has longitudinal grooves with reddish lines or blotches. Leaves triangular with an arrow-shaped base, alternate, simple, toothed or lobed, 2–10 cm long, sometimes with a greyish, mealy surface. Flowers minute, round-clustered, greenish, with petals absent and usually 2–5 sepals, growing on a tall spike. Seeds black, lens-shaped. Grows in disturbed sites, roadsides and cultivated fields across the prairies.

WARNING: *Saponins in the seeds are potentially toxic but are generally not well absorbed when ingested. The plant also contains some oxalic acid, an antinutrient that can block absorption of certain nutrients. However, this acid is neutralized by cooking.*

Bunchberry *Cornus canadensis*

Also called: Canada dogwood, dwarf dogwood • *C. unalaschensis*

Bunchberry (*C. canadensis*)

The bright scarlet or orange fruits of this woodland plant look like they should be very poisonous but are actually quite edible. However, opinions of their flavour range from insipid to a pulpy, sweet, flavourful fruit. The berries are abundant where the plant grows and were eaten by most First Nations within the plant's range; some groups only ate them as a snack, while others gathered large quantities and stored them for winter. They can be eaten raw as a trail nibble and are also said to be good cooked in puddings. However, each drupe contains a hard seed so be wary of mature dental work. Bunchberries (often mixed with other fruits) can be used whole to make sauces and preserves or cooked and strained to make beautiful scarlet-coloured syrups and jellies.

Bunchberry is reported to have anti-inflammatory, fever-reducing and pain-killing properties (rather like mild aspirin), but without the stomach irritation and potential allergic effects of salicylates. This plant also has a history of being used to treat headaches, fevers, diarrhea, dysentery and inflammation of the stomach or large intestine. The berries were eaten and/or applied in poultices to reduce the potency of poisons. They were also chewed and the resulting pulp applied topically to soothe and treat burns. Bunchberries have historically been steeped in hot water to make a medicinal tea for treating paralysis, or boiled with tannin-rich plants (such as common bearberry or commercial tea) to make a wash for relieving bee stings and poison-ivy rash. Native peoples used tea made with the entire plant to treat aches and pains, lung and kidney problems, coughs, fevers and fits. One Cree name for bunchberry, kawiskowimin, means "itchy chin berry," which refers to the itchy feeling resulting from rubbing the leaves or berries on the skin.

EDIBILITY: edible

FRUIT: Bright orange or red berry-like drupes, 6–9 mm wide, growing in dense clusters at the stem tips, nestled into a whorl of leaves (hence the common name bunchberry). The drupe has yellowish pulp and a single seed.

SEASON: Flowers May to July. Fruits ripen July to August.

DESCRIPTION: Perennial, rhizomatous herb, 5–20 cm tall. Leaves form a whorl appearance, 4–7, wintergreen-like, 2–8 cm long, with prominent veins. Flowers tiny, in a dense clump at the centre of 4 white to purple-tinged, petal-like bracts (exactly like miniature flowers of the dogwood tree), forming single, flower-like clusters about 3 cm across. Grows in cool moist woods and damp clearings at low to subalpine elevations, commonly found on rotting stumps and logs across the prairie provinces.

AMAZING: *This plant spreads its relatively heavy pollen grains through an interesting "explosive pollination mechanism." When the pollen is ripe and ready to be released, an antenna-like trigger lets go in the flower, rapidly springing the four pollen-laden anthers violently upwards together in a snapping motion, thereby catapulting the pollen grains far up into the air for dispersal.*

Clintonia & Queen's Cup *Clintonia* spp.

Queen's cup (*C. uniflora*)

The berry of this species, though very pretty and unusual looking, is dry, tasteless and mildly toxic to humans so it is not recommended for eating. The fruits make a good blue-coloured dye, though. The young leaves can be eaten raw or cooked and are said to taste slightly sweetish, like cucumber.

These plants, which spread through a network of underground rhizomes, mature to form a pretty colony of light green leaves with delicate flowers followed by metallic-blue berries. They are well worth growing for their ornamental value and are particularly striking when planted with other low-growing natives like bunchberry (white flowers and bright red berry clusters) and trillium (striking white to pink blooms). Native birds relish the fruit.

EDIBILITY: edible with caution (toxic)

FRUIT: Shiny, dark blue metallic-looking "bead" to 9 mm, growing atop a single stem.

SEASON: Flowers May to June. Fruits ripen July to August.

DESCRIPTION: Spreading perennial herbs arising from rhizomes, to 15–30 cm tall and often forming colonies. Leaves basal, 2–3 on each plant, 2–5 cm wide, to 30 cm long, shiny, narrowed at both ends. Flowers single, very occasionally double. Flowering stem slender, 14–40 cm tall,

FLOWERING PLANTS

Yellow clintonia (*C. borealis*)

erect, usually hairy at the top. Berries single, rarely double, round to oblong, coloured bright metallic blue, 6–12 mm thick. Found in shaded, moist to mesic forests and open woods.

Yellow clintonia (*C. borealis*) bears 3–8 yellow flowers. Found from MB eastward. Also called: yellow bluebead lily.

Queen's cup (*C. uniflora*) bears 1 or, rarely, 2 white flowers. Found in western AB (Rocky Mountains). Also called: one-flowered clintonia.

Yellow clintonia (*C. borealis*)

Queen's cup (*C. uniflora*)

Yellow clintonia (*C. borealis*)

Queen's cup (*C. uniflora*)

Fairybells *Prosartes* spp.

Rough-fruited fairybells (*P. trachycarpa*)

The berries of this species were not widely eaten by First Nations, and many tribes considered them poisonous. Some reports describe the fruit of Hooker's fairybells as somewhat sweet-tasting and juicy but growing so sparsely as to not warrant the effort to gather them. The fruit of the rough-fruited fairybells has been described as distinctly apricot-flavoured. Reports vary greatly as to whether these berries are edible or not, so caution is advised.

The fruits were sometimes taken to treat kidney problems, and the fresh seeds were used to clear matter from the eye or to treat snowblindness. Fairybells were associated with ghosts or snakes by some First Nations, and their names for these fruit include "snake berries" and "grizzly bear's favourite food." Rodents and grouse are known to feed on the berries. The leaves of this species are "drip tips," a form that ingeniously channels rainwater to the base of the plant.

EDIBILITY: edible with caution (toxic)

FRUIT: Berries egg-shaped, orange or yellow to bright red.

SEASON: Flowers April to July. Fruits ripen in July.

DESCRIPTION: Perennial herbs, 20–60 cm tall, with few branches, from thick-spreading rhizomes. Leaves alternate, broadly oval, 3–8 cm long, pointed at tips, rounded to a heart-shaped

FLOWERING PLANTS

base, fringed with short, spreading hairs, prominently parallel-veined. Flowers creamy to greenish white, narrowly bell-shaped, 10–20 mm long, drooping, 1–4 at branch tips. Grows in moist woods, forests and thickets as well as subalpine meadows.

Hooker's fairybells (*P. hookeri*) has smooth berries with 4–6 seeds and grows in southwest AB. Also called: Oregon fairybells • *Disporum hookeri*.

Rough-fruited fairybells (*P. trachycarpa*) has stamens that hang well below the flower petals, and conspicuously rough-skinned, velvety-surfaced orange to red berries with 6–12 seeds. Grows in all prairie provinces.
Also called: *Disporum trachycarpum*.

Rough-fruited fairybells (*P. trachycarpa*)

Rough-fruited fairybells (*P. trachycarpa*)

Hooker's fairybells (*P. hookeri*)

Hooker's fairybells (*P. hookeri*)

Hooker's fairybells (*P. hookeri*)

Twisted-stalks *Streptopus* spp.

Rosy twisted-stalk (*S. lanceolatus*)

These perennials are called "twisted stalks" because of the kinks (sometimes right-angled, sometimes just a sharp curve) present in the main stem or flower stalks. Most Native peoples regarded twisted-stalks as poisonous and used the plant mainly for medicine, but some tribes ate young plants and/or the bright-coloured berries, either raw or cooked in soups and stews. The berries are juicy and moderately sweet-tasting but are mildly toxic so should only be eaten in small quantities. Indeed, eating more than a few reportedly causes diarrhea, and it is best to consider these berries inedible.

Twisted-stalks were highly regarded for their general restorative qualities and were taken as a tonic or to treat general sickness. The whole plant was used by some First Nations to treat coughs, loss of appetite, stomachaches, spitting up blood, kidney trouble and gonorrhea. The blossoms were ingested to induce sweating. First Nations' names for the berries included owl berries, witch berries, black bear berries and frog berries; the berries were also believed to be eaten by snakes, deer and wolves. The plant was sometimes tied to, and used to scent, the body, clothes or hair.

Clasping twisted-stalk (*S. amplexifolius*)

The characteristic branched stem (sometimes zig-zagging) is what separates clasping twisted-stalk from the other twisted-stalks. Twisted-stalks are also different from the closely related fairybells in that the flowers attach to the stem in the leaf axils instead of to the branch tip.

EDIBILITY: edible with caution (toxic)

FRUIT: Berries hanging, red-orange or yellowish, egg-shaped and somewhat translucent; seeds small, whitish, somewhat visible.

SEASON: Flowers late June to early August. Fruits ripen August to September.

DESCRIPTION: Slender, herbaceous perennials from thick, short rhizomes, 0.4–1 m tall or more. Leaves smooth-edged, elliptical/oval-shaped, alternate, markedly parallel-veined. Flowers small, white, bell-shaped, 8–12 mm long, with 6 petals that flare backward, each hanging on the lower side of each stalk, attached to a stem at the leaf axil,

Rosy twisted-stalk (*S. lanceolatus*)

Twisted-stalks

1 per leaf. Found in moist shaded forests, clearings, meadows, disturbed sites and on streambanks at low to subalpine elevations.

Clasping twisted-stalk (*S. amplexifolius*) grows 0.5–1 m tall and has branched, smooth stems, sometimes bent at nodes, giving it a zig-zag appearance. Leaves clasping at base. Flowers greenish white. Berries yellow to bright red. Inhabits all prairie provinces.

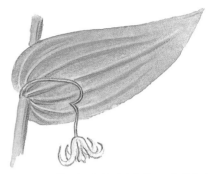

Clasping twisted-stalk (*S. amplexifolius*)

Rosy twisted-stalk (*S. lanceolatus*) grows to 30 cm tall, with stems usually unbranched, curved (not zig-zagged), leaves not clasping. Rose-purple or pink flowers with white tips; red berries. Grows in AB and MB. Also called: *S. roseus*.

Rosy twisted-stalk (*S. lanceolatus*)

FLOWERING PLANTS

WARNING: *Young twisted-stalk plants closely resemble green false-hellebore, which is extremely poisonous. Collecting the young shoots of twisted-stalks for consumption is not recommended unless you are absolutely sure of plant identification!*

Rosy twisted-stalk (*S. lanceolatus*)

Clasping twisted-stalk (*S. amplexifolius*)

Clasping twisted-stalk (*S. amplexifolius*)

139

Wild Lily-of-the-valley   *Maianthemum canadense*
Also called: Canada mayflower

Wild lily-of-the-valley (*M. canadense*)

Berries of this species are considered edible but are bitter-tasting and not very palatable to modern tastes. Caution is advised when eating berries of wild lily-of-the-valley because consuming too many can cause severe diarrhea. Wild lily-of-the-valley was eaten by some First Nations, but it was rarely highly regarded. The berries were usually eaten only casually by children or hunters and berry pickers while out on trips. Berries were generally eaten fresh when ripe.

The genus name is derived from the Latin words *maius*, meaning "May," and *anthemon*, meaning "flower," and refers to the flowering time of these plants (although, northern populations will flower later). The common name "wild lily-of-the-valley" comes from this plant's resemblance to the European lily-of-the-valley, *Convallaria majalis*. The fruit of this species is a true "berry" in botanical terms. These are lovely plants to look at and they are a good choice to plant in shady areas of the ornamental garden.

EDIBILITY: not palatable

FRUIT: Berries pea-sized, at first hard and green, ripening to cream-coloured with red speckles, then pinkish with red flecks, and finally to a solid red.

SEASON: Flowers May to June. Fruits ripen July to September.

FLOWERING PLANTS

DESCRIPTION: Herbaceous creeping perennial herb to 25 cm tall, arising from rhizomes and usually forming large colonies. Leaves stalkless, heart-shaped, alternating, usually 2 or 3, with prominent parallel veins. Flowers small, white, with 4 petals, 4–6 mm wide, borne in distinct terminal clusters. Fruit borne at the top of stems, in clusters. Found in moist woods and clearings across the prairie provinces.

WILD GARDENING: *This species grows into a delightful, dense carpet of delicate heart-shaped leaves and makes an excellent low-maintenance understorey planting for the woodland garden. The flowers are pretty in spring, and the berries provide a showy late-summer and fall display that attracts wildlife such as grouse that like to eat them.*

False Solomon's-seals *Maianthemum* spp.

False Solomon's-seal (*M. racemosum*)

The ripe berries, young greens and fleshy rhizomes of this species were eaten by various indigenous peoples across Canada. In cases where berries were eaten, it was usually casually (hunters, berry pickers, children). Berries of star-flowered false Solomon's-seal are said to be high in vitamin C. Some people consider the berries to be inedible, and they can have laxative effects, especially eaten fresh. Cooking the berries is said to lessen the laxative effect and improve the taste.

False Solomon's-seal was often combined with other plants for medicinal purposes. When combined with dogbane, it was used to keep the kidneys open during pregnancy, to cure sore throats and headaches and as a reviver. When mixed with black ash, the plant was used to loosen the bowels. When mixed with sweetflag, false Solomon's-seal was used as a conjurer's root to perform tricks or cast spells. Both the leaf and root were used to reduce bleeding. The rhizome was burned and inhaled to treat a number of ailments: to treat headaches, to quiet a crying child and to return someone to normal after temporary insanity. Leaf decoctions were applied to assist in childbirth and to treat itchy rashes. A rhizome decoction was used to treat back pain and overexertion.

EDIBILITY: not palatable

FRUIT: See individual species descriptions.

SEASON: Flowers May to June. Fruits ripen July to August.

DESCRIPTION: Tall, herbaceous perennials growing from thick, whitish, branching rhizomes, often found in dense clusters. Leaves smooth-edged, broad, elliptical, alternate along stems in 2 rows, 5–15 cm long, distinctly parallel-veined, often clasping. Flowers small, cream-coloured, 6-parted in dense,

False Solomon's-seal (*M. racemosum*)

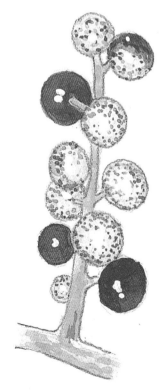

False Solomon's-seal (*M. racemosum*)

Star-flowered false Solomon's-seal (*M. stellatum*)

False Solomon's-seals

terminal clusters. Berries small and densely clustered, initially green and mottled or striped. Grows in woods, thickets, riverbanks and moist clearings (but see *M. trifolium*) across the prairie provinces.

False Solomon's-seal (*M. racemosum*) grows in clumps to 1.2 m tall from a fleshy, stout rootstock. Stems unbranching, arching. Flowers in clusters of 50–70. Berries a tight cluster of many seedy fruit. Berries at first green with copper spots, ripening to red, often with purple dots. Also called: feathery false lily-of-the-valley • *Smilacina racemosa*.

Star-flowered false Solomon's-seal (*M. stellatum*) grows 20–50 cm tall, flowers star-like, in clusters of 5–6, berries at first green with red stripes, dark blue or reddish black when ripe. Differentiated from *M. racemosum* by being smaller, having fewer flowers and leaves, and much fewer berries (2–8) that are larger, and green with red stripes when unripe. Also called: starry false lily-of-the-valley • *Smilacina stellata*.

Star-flowered false Solomon's-seal (*M. stellatum*)

Three-leaved Solomon's-seal (*M. trifolium*) grows 5–20 cm tall. Leaves oblong-lanceolate, 5–12 cm long and 2–4 cm wide, 2–4 (usually 3) on stem. Flowers 5–15, on a 5–10 cm panicle. Berries 4–6 mm, with 1–3 seeds, green with fine red dots when immature, ripening to red. Found in undisturbed peatlands, sphagnum bogs, muskeg, swamps and wet sites, but not in standing water in all prairie provinces. Also called: three-leaved false lily-of-the-valley • *Smilacina trifolia*.

False Solomon's-seal (*M. racemosum*)

FLOWERING PLANTS

Star-flowered false Solomon's-seal (*M. stellatum*)

Three-leaved Solomon's-seal (*M. trifolium*)

Star-flowered false Solomon's-seal (*M. stellatum*)

Smooth Solomon's-seal *Polygonatum biflorum*
Also called: great Solomon's-seal

Smooth Solomon's-seal (*P. biflorum*)

The berries of this species are considered toxic and inedible, and there is little history of its use. Some First Nations used the stems, leaves and roots for food. The dried powder of the roots were used for bread or boiled, and the leaves and stems were cooked, sometimes fried in grease.

The rhizomes have a history of medicinal use in the treatment of skin discoloration, ulcers and broken bones. The berries are said to taste mucilaginous, sweet, then acrid.

This is a very decorative and charming plant, with its delicate arching stems, demure blooms, striking berries and fine foliage that turns a lovely yellow in fall. Smooth Solomon's-seal makes a great addition to the shade or woodland garden, and once its tough rhizomes are established, it slowly spreads to create a drought-resistant patch. This plant is a great attraction for hummingbirds, which feed on the flower nectar.

EDIBILITY: not palatable, edible with caution (toxic)

FRUIT: Blue-black berries to 1.5 cm in diameter.

SEASON: Flowers April to June. Fruits ripen July to August.

DESCRIPTION: Herbaceous perennial growing 0.3–1.5 m tall and generally 45–60 cm wide from a thick, fleshy, many-jointed white rhizome. Often spreads into colonies. Leaves linear to ovate, simple, finely veined, alternate, opposite in two close ranks. Flowers bell-shaped, tubular, greenish white, 1.5–3 cm long, borne below arching stems, normally in clusters of 2–5. Inhabits forests and woodlands in MB and SK, preferring shade or partial sun in light, well-drained, humus-rich soil.

Strawberries *Fragaria* spp.

Wood strawberry (*F. vesca*)

These delicious little berries pack significantly more flavour than a typical large, domestic strawberry. Wild strawberries are small compared to modern cultivars and are probably best enjoyed as a nibble along the trail, but they can also be collected for use in desserts and beverages. A handful of bruised berries or leaves, steeped in hot water, makes a delicious tea served either hot or cold.

Strawberries were a popular berry with all prairie First Nations, but their juiciness can make them difficult to dry and preserve. Today, strawberries are preserved by freezing, canning or making jam, but traditionally they were sun- or fire-dried. The berries were mashed and spread over grass or mats to dry into cakes, which were later eaten or rehydrated, either alone or mixed with other foods as a sweetener. Anyone who has had the extreme pleasure of savouring dried wild strawberries knows that this is a treat well worth the time to prepare! Strawberry flowers, leaves and stems were also sometimes mixed with roots of other plants in cooking pits as a flavouring.

Strawberries contain many quickly assimilated minerals (e.g., sodium, calcium, potassium, iron, sulphur and silicon), as well as citric and malic acids, and they were traditionally prescribed to enrich the bloodstream.

Strawberries are a good source of ellagic acid, a chemical that is believed to prevent cancer. To remove tartar and whiten discoloured teeth, strawberry juice can be held in the mouth for a few minutes and then rinsed off with warm water. This treatment is reported to be most effective with a pinch of baking soda in the water. Large amounts of fruit in the diet also appear to slow dental plaque formation. Strawberry juice, rubbed into the skin and later rinsed off with warm water, has been used to soothe and heal sunburn. The berries were also used as a deodorant.

Wild strawberry (*F. virginiana*)

Wild strawberry (*F. virginiana*)

Many people will be surprised to learn that the strawberry is technically not a fruit! What we think of as the "berry" is actually a swollen receptacle (this is the base of the flower, which you would normally expect to see inside a fruit). The true "fruits" are the tiny dark seeds (seed-like achenes) that are easily found either embedded in, or perched on, the soft flesh of the strawberry.

EDIBILITY: highly edible

FRUIT: Berries are red when ripe, resembling miniature cultivated strawberries.

SEASON: Flowers May to August. Fruits ripen starting in June, and flowers continue to bloom throughout the season, so plants often have ripe fruit and flowers on them at the same time.

DESCRIPTION: Low-creeping perennials with long slender runners (stolons). Leaves green, often turning red in fall, 5–10 cm across, with 3 sharply toothed leaflets. Flowers white, 5-petalled, 1.5–2 cm across, usually several per stem, forming

Strawberries

Wood strawberry (*F. vesca*)

Wood strawberry (*F. vesca*)

Wild strawberry (*F. virginiana*)

FLOWERING PLANTS

small, loose clusters. Found in open woods and meadows across the prairies.

Wood strawberry (*F. vesca*) has yellowish green leaflets, with the end tooth projecting beyond its adjacent teeth. Leaflets are thick, silky hairy (mostly below), strongly veined and scalloped. The stalks of the flowers and fruit are longer than the leaves.

Wild strawberry (*F. virginiana*) has bluish green leaflets, hairless, with the end tooth narrower and shorter than its adjacent teeth. Flowers and fruit are borne on short stalks, usually just at ground level. Also called: blueleaf strawberry.

Wild strawberry (*F. virginiana*)

Wild strawberry (*F. virginiana*)

Wild Berry Muffins

Makes 12 muffins

This batter can also be baked in a loaf form.

5 Tbsp vegetable oil • 2 eggs, lightly beaten
1½ cups mixed wild berries (strawberries, thimbleberries, blueberries, huckleberries, etc.)
1 tsp salt • 1¾ cups whole-wheat flour
¾ cup brown sugar • 2¼ tsp baking powder

Preheat oven to 400° F. Mix wet ingredients together in a bowl. Sift dry ingredients together in another bowl. Make a shallow well in the centre of the dry ingredients and slowly add the wet mixture. Mix well and pour into greased or lined muffin tins. Bake for 10 to 15 minutes, or until a knife inserted into a muffin comes out clean.

Northern Comandra   *Geocaulon lividum*
Also called: false toadflax • *Comandra livida*

Northern comandra (*G. lividum*)

The berries of this Canada-wide species are edible, but reports vary considerably regarding their tastiness. Northern comandra fruit were eaten by the Fisherman Slave Lake tribe but were considered inedible by most First Nations within the plant's range and are probably best used as an emergency food if lost in the woods. The Chipewyan considered the berries inedible but a few berries were eaten once a year to treat persistent chest trouble.

This species is parasitic, feeding off the roots of host plants (e.g., strawberries, cranberries, bearberries, asters). The species attaches itself to the roots of other plants through sucker-like organs on their rootlets. They are able to get additional nutrients this way.

EDIBILITY: not palatable

FRUIT: Fleshy orange to scarlet berry-like drupe, 6–10 mm.

SEASON: Flowers May to June. Fruits ripen August to September.

DESCRIPTION: Perennial deciduous herb from reddish, creeping underground rhizomes, growing 10–25 cm tall. Flowering stems erect, unbranched. Leaves simple, alternate, short-stalked, oval, 1–2.5 cm long and 0.5–1 cm wide, pale grey to purplish tinged, toothless margins. Flowers grow in clusters of 2–4 from leaf axils, green to purplish, 5 petal-like sepals, inconspicuous. Fruits single, rarely double. Inhabits moist coniferous forests, woods, and sphagnum bogs across the prairie provinces.

Western Poison-ivy   *Toxicodendron rydbergii*

Also called: *Rhus radicans* var. *rydbergii*

Western poison-ivy (*T. rydbergii*)

Poison-ivy plants contain an oily resin (urushiol) that causes a nasty skin reaction in most people, especially on sensitive skin and mucous membranes. Since the allergic contact dermatitis appears with some delay after exposure, many people do not realize they have come into contact with these plants until it is too late. Sensitization can also lead to more severe reaction after repeated exposure.

Urushiol is not volatile and therefore it is not transmitted through the air, but it is persistent and can be carried to unsuspecting victims by pets, through clothing and tools and even on smoke particles from burning poison-ivy plants. The resin can persist on pets and clothing for months and is also ejected in fine droplets into the air when the plants are pulled.

Washing with a strong soap can remove the resin and prevent a reaction if it is done shortly after contact. Washing also prevents transfer of the resin to other parts of the body or to other people. Be sure to use cold water as warm water can help

POISONOUS PLANTS

the resin penetrate the skin, where it is extremely difficult to remove. The liquid that oozes from poison-ivy blisters on affected skin does not contain the allergen. Ointments and even household ammonia can be used to relieve the itching of mild cases, but people with severe reactions might need to consult a doctor.

The scientific name for this species comes from the Latin *toxicum*, meaning "poison," and the Greek *dendron*, meaning "tree."

EDIBILITY: poisonous

FRUIT: Whitish to brown, berry-like drupes 4–5 mm wide.

SEASON: Flowers June to July. Fruits ripen July to August.

DESCRIPTION: Trailing to erect deciduous shrub, growing to 10–30 cm tall and forming colonies. Leaves alternate, bright glossy green, resinous, with irregular notches, made up of 3 strongly veined leaflets, scarlet in autumn. Flowers yellowish-green, 5-petalled, 1–3 mm across, forming clusters. Separate male and female plants.

Devil's Club *Oplopanax horridus*
Also called: *Echinopanax horridum*

Devil's club (*O. horridus*)

Throughout its range, devil's club (which is botanically related to the ginseng family) is considered to be one of the most powerful and important of all medicinal plants. The berries of this plant are considered inedible, perhaps partly because they are held high above a remarkable fortress of irritating spiny leaves and stems and because even the berries have spikes! Although devil's club tea is recommended today for binge-eaters who are trying to lose weight, some tribes used it to improve appetite and to help people gain weight. In fact, it was said that a patient could gain too much weight if it was used for too long. Some tribes used a strong decoction of the plant to induce vomiting in purifying rituals preceding important events such as hunting or war expeditions. This decoction was also applied to wounds to combat staphylococcus infections, and ashes from burned stems were sometimes mixed with grease to make salves to heal swellings and weeping sores. The bright red berries were rubbed into hair to combat dandruff and lice and to add shine.

Like many members of the ginseng family, devil's club contains glycosides that are said to reduce metabolic stress and thus improve one's sense of well-being. The roots and bark of this plant contain the majority of active compounds and have traditionally been used in the treatment of arthritis, diabetes, rheumatism, digestive troubles, gonorrhea and ulcers. The root tea has been reported to stimulate the respiratory tract and to help bring

POISONOUS PLANTS

up phlegm when treating colds, bronchitis and pneumonia. It has been used to treat diabetes because it helps regulate blood sugar levels and reduces the craving for sugar. Indeed, devil's club extracts have successfully lowered blood sugar levels in laboratory animals.

Possibly because of its diabolical spines, devil's club was considered a highly powerful plant that could protect one from evil influences of many kinds. Devil's club sticks were used as protective charms, and charcoal from the burned plant was used to make protective face paint for dancers and others who were ritually vulnerable to evil influences.

EDIBILITY: poisonous

FRUIT: Bright red, berry-like drupes, slightly-flattened, sometimes spiny, 5–8 mm long, in showy pyramidal terminal clusters.

SEASON: Flowers May to July. Fruits ripen July to September.

DESCRIPTION: Strong-smelling, deciduous shrub, 1–3 m tall, with

spiny, erect or sprawling stems. Leaves broadly maple leaf–like, 10–40 cm wide, with prickly ribs and long, bristly stalks. Spines on leaves and stems grow up to 9 mm long. Flowers greenish white, 5–6 mm long, 5-petalled, forming 10–25 cm-long, pyramid-shaped clusters of bright red berries. Grows in moist, shady foothill and montane sites in AB. Easy to find when you lose your footing on a rough trail as it is invariably the plant that you grab onto to stop your fall.

WARNING: *Devil's club spines are brittle and break off easily, embedding in the skin and causing infection. Some people have an allergic reaction to the scratches from this plant. Wilted leaves can be toxic, so only fresh or completely dried leaves should be used to make a medicinal tea, but even then the tea should be taken under the guidance of a registered herbalist and in moderation because extended use can irritate the stomach and bowels.*

American Bittersweet *Celastrus scandens*

American bittersweet (*C. scandens*)

The showy, berry-like fruits of American bittersweet provide winter food for wildlife species such as grouse, pheasant, quail, rabbit and squirrel. To humans, however, they are quite poisonous. First Nations reportedly boiled the bark or inner bark of the stem, said to be sweet and palatable, to make a thick soup. However, the bark was considered a starvation food and eaten only when other foods were in short supply.

Although this is a medicinal plant with a history of use by First Nations, it is scarcely used today by modern herbalists. Traditionally, the root bark tea was used to induce sweating, urination and vomiting and was considered effective in the treatment of chronic liver and skin disorders, rheumatism, leucorrhea, dysentery and irregular menstruation. The tea was also given to pregnant women alone or combined with red raspberry leaves to reduce the pain of childbirth. An infusion of the leaves and stems was given to pregnant women to reduce fever and soreness. An infusion of the bark was taken for upset stomachs, and a decoction was taken for bowel complaints and as a physic, especially for babies. An infusion of the root was prescribed for kidney trouble following childbirth and a leaf tea for diarrhea and dysentery. A decoction of the roots was used as a wash on the lips of children who had misbehaved. The root was chewed to treat coughs and tuberculosis. The chewed root was also traditionally smeared on the body, believing that the user would be impervious to wounding. As a poultice, the boiled root was used to treat obstinate sores and other skin eruptions. The bark was applied externally as an ointment on burns, scrapes, tumours and skin eruptions. Extracts of the bark are reportedly cardioactive.

The most important use today for bittersweet is as a decorative vine, available in farmer's markets and nurseries from spring to autumn. Related plants are under study as botanical insecticides.

EDIBILITY: poisonous

FRUIT: Fruits yellow orange, opening when ripe and exposing the scarlet, berry-like interior.

SEASON: Flowers May to June. Fruits ripen August to September.

DESCRIPTION: Twining, woody vine with alternate, oval to lance-shaped, finely serrated leaves, 5–10 cm long. Flowers small, 4 mm wide, green with 4–5 petals, in terminal clusters to 10 cm long. Grows in thickets, woods and on riverbanks in southern SK and MB.

Snowberries *Symphoricarpos* spp.

See also: creeping snowberry • *Gautheria hispidula* (p. 101)

Common snowberry (*S. albus*)

Some sources report that snowberries are edible, though not very good. However, they are toxic and in large quantities can be poisonous. Most tribes considered snowberries poisonous and did not eat them. Some believed they were the ghosts of saskatoons, part of the spirit world and not to be eaten by the living. The fruits were used as a wash to treat sore eyes. The berries were also given to horses to treat water retention.

The spongy white berries are fun to squish and pop—rather like bubble wrap! The unusual white berries persist on the plant through winter, providing a showy and decorative display that in mild winters can last well into spring. The berries make a wonderful addition to winter holiday wreaths, garlands and other festive decorations.

This is a very drought-tolerant and decorative species that will thrive on steep or unstable slopes and other areas that may otherwise be difficult to

Common snowberry (*Symphoricarpos albus*)

landscape. The leaves and flowers, albeit small, are pretty, and the white berries provide winter forage for birds and small mammals.

EDIBILITY: edible with caution (toxic), poisonous

FRUIT: White, waxy, spongy berry-like drupes, 6–10 mm long, singly or in clusters on stem tips, often lasting through winter.

SEASON: Flowers May to August. Berries ripen and whiten August to September.

DESCRIPTION: Erect, deciduous shrubs up to 1 m tall. Stems flexible and strong, grey in colour, with bark becoming shredded on more mature specimens. Leaves pale green, opposite. Flowers pink to white, broadly funnel-shaped, 4–7 mm long, borne in small clusters in upper leaf axils or stem tips. Spreads rapidly through a tough, dense, underground root system (rhizomes) and quickly forms an impenetrable thicket if left alone. Grows on rocky banks, hedgerows, forest edges and roadsides across the prairies.

Common snowberry (*S. albus*) has thin oval leaves, 2–4 cm long. Flowers with non-hairy style, which, along with stamens, do not protrude from the flower. Also called: thin-leaved snowberry.

Western snowberry (*S. occidentalis*) has thick oblong leaves, 3–6 cm long. Flowers with hairy style, which, along with stamens, usually protrudes from the flower. The fruit turns purplish in autumn. Also called: buckbrush.

> **WARNING:** *The branches, leaves and roots of snowberries are poisonous, containing the alkaloid chelidonine, which can cause vomiting, diarrhea, depression and sedation. The berries contain several flavonoid glycosides, and when consumed can cause symptoms that include vomiting, dizziness and delirium, blood-stained urine, sedation and even death at high doses.*

Western snowberry (*S. occidentalis*)

Western snowberry (*S. occidentalis*)

Honeysuckles *Lonicera* spp.

Twinflower honeysuckle (*L. involucrata*)

Although there are some accounts of First Nations eating honeysuckle berries occasionally for food, most considered them to be inedible or poisonous. The flowers, however, produce sweet nectar at their base that can be sucked out. This treat was especially enjoyed by children.

Honeysuckle was employed by a number of First Nations for a diverse range of ailments. The berries were used as a cathartic and emetic to cleanse the body and to treat stomach and chest troubles as well as sore eyes and dandruff. An infusion of the bark of twining honeysuckle was used as a cathartic and diuretic to treat kidney stones, menstrual difficulties and dysuria. A tea made from the peeled, internodal stems was used for urine retention, flu and blood clotting after childbirth. A decoction of the root was mixed with other ingredients to treat gonorrhea. Pregnant women ingested a tea made from the berries and bark to expel worms. A decoction of the whole plant was given to children for fevers and general sickness.

Twinflower honeysuckle bark was taken for coughs, and the leaves were chewed and applied externally to itchy skin, boils and gonorrhoeal sores. A decoction of twinflower leaves or inner bark was used as daily eyewash to bathe sore eyes. Boiled bark was applied to burns, infections and wounds. In spite of the myriad traditional medical uses of honeysuckle, however, the plant is rarely used today in modern herbalism.

The stems, which are supple and relatively strong, were used by First Nations as building materials and to make fibres for mats, baskets, bags, blankets and toys. Children used the hollow stems as straws. Fruit of

POISONOUS PLANTS

twinflower honeysuckle were used as a purple dye for basketry materials and for painting faces on dolls and other items. The plant was used in a number of love potions, charms and medicines to either form or destroy a relationship. During Victorian times, teenage girls were told not to bring honeysuckle home in the belief that the flowers induced erotic dreams. Ravens, crows and bears are said to eat honeysuckle berries in large quantities.

EDIBILITY: not palatable, poisonous

FRUIT: An orange, red, blue or black berry to about 1 cm across containing several seeds.

SEASON: Flowers May to July. Fruits ripen July to September.

DESCRIPTION: Twining, deciduous woody vines or shrubs, trailing or climbing or growing to 3 m or more tall, with opposite, simple leaves, to 10 cm long. Flowers sweet-scented, bell-shaped, with a sweet, edible nectar.

Twining honeysuckle (*L. dioica*) is a twining, woody vine growing to 3 m. Young stems green to purplish/red, older stems grey/brown with shredding bark. Leaves, hairless, oval, uppermost leaves joined at base, forming cup around stem. Flowers yellow to orange (sometimes dark reddish with age), 2–2.5 cm long, growing clustered on a short stem from the centre of a terminal cup. Berries smooth, oval, reddish, growing in a cluster at the stem tips. Fruit are extremely bitter and not recommended for eating. Grows in dry woods, thickets and rocky slopes across the prairie provinces. Also called: limber honeysuckle, red honeysuckle.

Twinflower honeysuckle (*L. involucrata*) grows to 1–3 m tall, with 4-angled twigs that are greenish when young, greyish with shredding bark when older. Leaves broadly lance-shaped, to 5–10 cm long, with gland-tipped hairs, sharp-pointed at tip. Flowers bell-shaped, yellow, 1–2 cm long, in pairs in leaf axils, surrounded by fused bracts (an "involucre," hence the scientific name *involucrata*). Fruits shiny, black berries, to 1.2 cm across, fused bracts turn from green to deep purplish maroon as fruits ripen. Berries are bitter-tasting and reported as inedible to poisonous. Grows in moist or wet soil in forests, clearings, riverbanks, swamps and thickets across the prairies. Also called: bracted honeysuckle, black twinberry.

Twinflower honeysuckle (*L. involucrata*)

Twining honeysuckle (*L. dioica*)

Canada Moonseed *Menispermum canadense*

Canada moonseed (*M. canadense*)

The fruit of this species contains dauricine, an alkaloid that affects "excitable" muscles such as the heart and smooth muscles of blood vessels. By blocking calcium flow, dauricine can decrease blood pressure and control the heart rate. A tincture of moonseed root was historically prescribed in the treatment of scrofula, rheumatism, gout and skin diseases.

EDIBILITY: poisonous

FRUIT: Purplish to black drupes, 6–7 mm, growing in loose grape-like clusters, containing a single crescent-shaped seed.

SEASON: Flowers May to July. Fruits ripen August to September.

DESCRIPTION: Climbing perennial vine to 7 m. Stems thin, twining, dark maroon, woody below. Young stems greenish or reddish, flexible, hairy. Climbs by twining around other plants

POISONOUS PLANTS

WARNING: *Do not confuse this species with riverbank grapes, a vine that is related, but with only a superficially similar appearance. The dark blue, berry-like fruits of moonseed are highly poisonous and may be mistaken for edible grapes, particularly by children. The leaves of moonseed are smooth rather than toothed, and each fruit contains a single crescent-shaped seed rather than the many seeds found in riverbank grapes.*

and structures and has no tendrils. Leaves alternate, 3–7 broad, shallow lobes, smooth margins, upper surface hairy, lower leaf surface silvery green and slightly hairy. Inhabits sunny areas in moist woods, streambanks, hedges and thickets in southern MB.

Red Baneberry   *Actaea rubra*
Also called: common baneberry, snake berry • *A. arguta, A. eburnea*

Red baneberry (*A. rubra*)

Baneberry is related to the commercial phytomedicine black cohosh, and some indigenous peoples used baneberry root tea in a similar way to treat menstrual and postpartum problems, as well as colds, coughs, rheumatism, wounds, nosebleeds and syphilis. Herbalists have used baneberry roots as a strong antispasmodic, anti-inflammatory, vasodilator and sedative, usually for treating menstrual cramps and menopausal discomforts.

Baneberry is a striking-looking plant with attractive foliage and delicate stems of puffy white flowers in spring followed by showy spikes of red or white berries in fall. Planted with ferns, hostas and other shade-loving species, it makes for a decorative addition to the shade garden. Many small birds and animals eat these berries.

EDIBILITY: poisonous

FRUIT: Very showy, several-seeded, glossy red or white berries 6–8 mm long, growing singly on a long stalk.

POISONOUS PLANTS

Season: Flowers May to July. Fruits ripen July to August.

Description: Branched, leafy, generally solitary perennial herb, 30 cm–1 m tall, from a woody stem base and fibrous roots. Stems long, wiry. Leaves are coarsely toothed, alternate, few and large, divided 2–3 times in threes, crowded at base of stem and sparser near the top. Flowers white, with 5–10 slender, 2–3 mm long petals, forming long-stalked, cone-shaped clusters. Inhabits deciduous and mixed coniferous forests, subalpine meadows, moist woodlands, streambanks and swamps at low to montane elevations in all prairie provinces.

Warning: *All parts of baneberry are poisonous, but the roots and berries are most toxic. Indeed, the common name "baneberry" derives from the Anglo-Saxon* bana, *which means "murderous." Eating as few as 2–6 berries can cause severe cramps and burning in the stomach, vomiting, bloody diarrhea, increased pulse, headaches and/or dizziness. Acute poisoning results in convulsions, paralysis of the respiratory system and cardiac arrest. No deaths have been reported in North America, probably because the berries are extremely bitter and unpleasant to eat.*

Nightshades *Solanum* spp.

Also called: bittersweets, matrimony vines

European bittersweet (*S. dulcamara*)

The immature (green) berries and leaves of European bittersweet contain toxic alkaloids that can cause vomiting, dizziness, a weakened heart, liver damage, convulsions, paralysis and even death. Cattle and sheep have been poisoned by eating these bitter plants, but livestock deaths are rare. The ripe berries contain only small amounts of alkaloids and are not considered a threat if eaten in moderation, but large amounts could prove toxic.

Stem extracts of European bittersweet have been taken internally as a sedative and pain reliever, for increasing urination and for treating asthma, gout, rheumatism, whooping cough and bronchitis, but pharmacological evidence does not support these uses. Extracts are reported to have antibiotic activity, which could be useful in salves and lotions for combating infection. These plants have been used for many years to treat skin diseases, sores, swellings and inflammations around fingernails and toenails. Recent research has shown that bittersweet contains beta-solanine, a tumour-inhibiting compound that has potential in treating cancer.

EDIBILITY: poisonous

FRUIT: Round to oblong berries, 8–11 mm wide.

SEASON: Flowers June to July. Fruits ripen August to September.

POISONOUS PLANTS

DESCRIPTION: Rhizomatous annuals or perennial herbs and vines, often somewhat woody; stems leafy, 1–3 m tall or long, erect or tending to climb, scramble or sprawl on other vegetation. Leaves alternate, oval and entire margined. Flowers with a yellow cone projecting from the centre of 5 back-curved, white to blue petals, borne in loose, branched clusters.

Black nightshade (*S. americanum* var. *nodiflorum*) is an erect annual, taprooted herb with white flowers and black berries. This species, introduced in our country from the southern and eastern U.S., grows in disturbed sites (roadsides, thickets) in all prairie provinces.

European bittersweet (*S. dulcamara*) is a perennial vine, trailing or climbing on other vegetation, with blue-violet to purple flowers and bright red berries. This introduced Eurasian species can be found in thickets and clearings near habitation in all prairie provinces. Also called: climbing nightshade, woody nightshade.

European bittersweet (*S. dulcamara*)

Black nightshade (*S. americanum*)

169

Pacific Yew *Taxus brevifolia*

Also called: western yew, mountain mahogany

Pacific yew (*T. brevifolia*)

The fleshy red part of the fruit is said to be edible, but the seeds are extremely poisonous, so this fruit is not recommended for consumption (see Warning). The leaves are also toxic. The bark of Pacific yew is an original source of the anticancer drug taxol. After a long period of development by the National Cancer Institute and pharmaceutical partners, this drug was approved for use in treating a variety of cancers and was particularly successful in treating breast and ovarian cancers. The slow-growing Pacific yew, however, became quickly depleted in the wild by unregulated over-harvesting for the pharmaceutical industry. There are ongoing concerns regarding its natural regeneration since it was also removed for many years as a "weedy" species in second-growth timber stands, and it requires both a male and female tree growing in relative proximity to each other to reproduce. Taxane derivatives are now in great part obtained from managed harvest of the more common Canada yew (*T. canadensis*) in eastern Canada, from which the drug is prepared by extraction and semi-synthesis. Some Native peoples used yew bark for treating illness (indeed, this is how modern researchers first knew to research this plant) and applied the wet needles as poultices on wounds, but do not try this remedy.

The heavy, fine-grained wood of Pacific yew is extremely durable and can endure great stress without breaking.

POISONOUS PLANTS

The wood was prized by First Nations within the plant's range to make items such as bows, wedges, clubs, paddles, digging sticks, fish hooks, tool handles and harpoons. Bows, carved from seasoned yew wood, were varnished with boiled animal sinew and muscle. Pacific yew was also used to make sewing needles, awls, dipnet frames, knives, dishes, spoons, boxes, dowels and pegs, canoe spreaders, bark scrapers, fire tongs, combs and snowshoe frames. The wood was a valuable trade item to exchange with other peoples where the tree does not naturally grow. Two ancient spears, dating from the early Stone Age, were found to be made of yew wood. The wood is still prized today by carvers, but it is relatively scarce because of the tree's slow growth and reproduction. With their dark, evergreen needles and scarlet berries, yews make lovely ornamental shrubs, but their poisonous seeds, bark, branches and leaves are dangerous to children and some animals.

EDIBILITY: edible with **extreme** caution (toxic), poisonous

FRUIT: Berry-like arils, 4–5 mm across, with a cup of orange to red fleshy tissue around the single bony seed. The showy berry-like fruit of this species, with its sweet taste but slimy texture, has historically been considered edible. However, the hard seeds found within the fleshy cup are extremely poisonous, so this fruit is not recommended for consumption (see Warning).

SEASON: Flowers in June. Berries ripen to an orange or deep red August to October.

DESCRIPTION: Small, generally scraggly-looking evergreen shrub or tree, with a straight trunk, to 15 m tall (rarely to 25 m). Branches drooping, bark reddish brown, scaly and flaking. Often grows together in small thickets. Needles soft, flattened, 3.5 cm long, arranged alternately in two rows, glossy green above and paler green underneath with two whitish bands of stomata and a sharp tip. Male and female are separate trees, the male pollen-bearing cones inconspicuous. The tiny green flowers of the female tree eventually produce scarlet-red arils. Unusually, this conifer has no pitch. Grows in moist, shady sites such as streambanks and under mature coniferous forests, at low to montane elevations in AB.

WARNING: The needles, bark and seeds contain **extremely** poisonous, heart-depressing alkaloids called taxanes. Drinking yew tea or eating as few as 50 leaves can cause death. The berries (which take 2 years to mature) are eaten by many birds, and the branches are said to be a preferred winter browse for moose, but many horses, cattle, sheep, goats, pigs and deer have been poisoned from eating yew shrubs, especially when the branches were previously cut.

False Virginia Creeper *Parthenocissus vitacea*
Also called: woodbine

False Virginia creeper (*P. vitacea*)

The berries (which contain high levels of oxalic acid) and foliage of this plant are considered poisonous. The Ojibway, however, reportedly consumed a very closely related species, Virginia creeper (*P. quinquefolia*). They boiled the stalks and ate the inner bark like corn on the cob. A syrup was also rendered by boiling the stalks and then used to cook wild rice. The Iroquois used a decoction of false Virginia creeper to treat urinary problems and skin ailments.

The most common modern use for this plant is as an attractive garden climber or groundcover, particularly for its showy autumn foliage, which provides a vibrant display of fiery red, purple and scarlet leaves after the first frosts. It is also notable for its purplish-black berries and red stems, which remain on the plant after the leaves have fallen, providing an interesting winter display.

False Virginia creeper is useful as a wildlife attractant. The berries are a popular food for winter birds, in particular many species of songbirds, and wild deer and livestock browse its foliage. Because of its perennial root system and ground-covering habit, this plant has been used in slope stabilization, and it can also provide valuable habitat for small mammals and birds.

POISONOUS PLANTS

EDIBILITY: poisonous

FRUIT: Round, purplish black berries, 8–10 mm in diameter, with a thin layer of flesh around 3–4 seeds; fruit stalks often bright red.

SEASON: Flowers June to July. Fruits ripen August to September.

DESCRIPTION: Woody, scrambling or climbing, deciduous vine reaching 20–30 m in length or height; tendrils sparsely branched. Leaves alternate, long-stalked, palmately compound with 5 leaflets; leaflets elliptic to oval, long-tapered at tips, wedge-shaped at base, 5–12 cm long, dark green and shiny above, paler beneath, short-stalked; margins coarsely and sharply toothed above middle; becoming brilliant scarlet in autumn. Flowers greenish, small, about 5 mm across; 25–200 or more in forked, branching clusters. Found in moist soils in woods and thickets and open ground along roadsides in southern MB.

COMPARE: This plant is almost indistinguishable from true Virginia creeper (P. quinquefolia), the only major difference being that false Virginia creeper lacks the adhesive, sticky discs on its tendrils that allow the true Virginia creeper to climb smooth surfaces. False Virginia creeper rambles through trees and along the ground, attaching itself with twining tendrils similar to those of a grape, rather than vertically adhering itself to walls, trellises and bare tree trunks. The foliage (including showy autumn colouring) of these 2 species is identical.

Glossary

accessory fruit: a fruit that develops from the thickened calyx of the flower rather than from the ovary (e.g., soapberry).

achene: a small, dry fruit that doesn't split open; often seed-like in appearance; distinguished from a nutlet by its relatively thin wall.

alkaloid: any of a group of bitter-tasting, usually mildly alkaline plant chemicals. Many alkaloids affect the nervous system.

alternate: situated singly at each node or joint (e.g., as leaves on a stem) or regularly between other organs (e.g., as stamens alternate with petals).

anaphylaxis: a hypersensitivity reaction to the ingestion or injection of a substance (a protein or drug) resulting from prior contact with a substance. Anaphylaxis can progress rapidly and be life-threatening.

annual: a plant that completes its life cycle in one growing season.

anthers: the pollen-producing sacs of stamens.

anthraquinone: an organic compound found in some plants that has a laxative effect when ingested. It is also used commercially as a dye and pigment, and also in the pulp and paper industry.

aril: a specialized cover attached to a mature seed.

armed: a plant furnished with defensive bristles or thorns.

axil: the position between a side organ (e.g., a leaf) and the part to which it is attached (e.g., a stem).

berry: a fleshy, simple fruit that contains one or more ovule-bearing structures (carpels) that each contain one or more seeds; the outside covering (endocarp) of a berry is generally soft, moist and fleshy (e.g., blueberry).

biennial: a plant that lives for two years, usually producing flowers and seed in the second year.

bitters: alcoholic drinks consumed with a meal which contain bitter herbs to aid in the process of digestion (e.g., Swedish bitters or Angostura bitters).

bog: a peat-covered wetland characterized by *Sphagnum* mosses, heath shrubs and sometimes trees.

bract: a specialized leaf with a flower (or sometimes a flower cluster) arising from its axil.

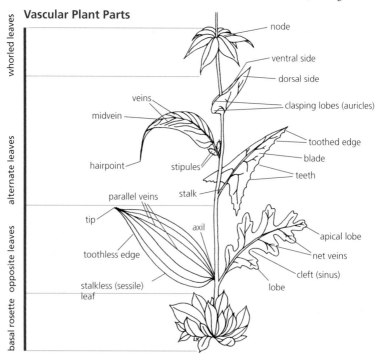

Vascular Plant Parts

GLOSSARY

calcareous: a type of soil with a high calcium content.

calyx: the outer (lowermost) circle of floral parts; composed of separate or fused lobes called sepals; usually green and leaf-like.

carpel: a fertile leaf bearing the undeveloped seed(s); one or more carpels join together to form a pistil.

cathartic: a substance that purges the bowels.

compound leaf: a leaf composed of two or more leaflets.

compound drupe: a collection of tiny fruit that forms within the same flower from individual ovaries; this type of fruit is often crunchy and seedy (e.g., boysenberries).

cone: a fruit that is made up of scales (sporophylls) that are arranged in a spiral or overlapping pattern around a central core, and in which the seeds develop between the scales (e.g., juniper).

corolla: the second circle of floral parts, composed of separate or fused lobes called petals; usually conspicuous in size and colour, but sometimes small or absent.

cultivar: a plant or animal originating in cultivation (e.g., loganberry or Golden Delicious apple).

deciduous: having structures (leaves, petals, seeds, etc.) that are shed at maturity and in autumn.

drupe: a fruit with an outer fleshy part covered by a thin skin and surrounding a hard or bony stone that encloses a single seed (e.g., a plum).

drupelet: a tiny drupe; part of an aggregate fruit such as a raspberry.

emetic: induces vomiting.

endocarp: the inner layer of the pericarp.

eulachon grease: grease from the eulachon (*Thaleichthys pacificus*), a small species of fish in the smelt family that lives most of its life in the Pacific Ocean but comes inland to fresh water for spawning.

fruit: a ripened ovary, together with any other structures that ripen with it as a unit.

glabrous: without hair, smooth.

glandular: associated with a gland (e.g., glandular hair).

glaucous: a frosted appearance due to a whitish powdery or waxy coating.

globose: shaped like a sphere.

glycoside: a two-parted molecule composed of a sugar and an aglycone, usually becoming poisonous when digested and the sugar is separated from its poisonous aglycone.

habitat: where a plant or animal is normally found; the characteristic environmental conditions in which a species is normally found.

haw: the fruit of a hawthorn, usually with a fleshy outer layer enclosing many dry seeds.

heath: a member of the heath family (Ericaceae).

herbaceous: a plant or plant part lacking lignified (woody) tissues.

hip: a fruit composed of a collection of bony seeds (achenes), each of which comes from a single pistil, covered by a fleshy receptacle that is contracted at the mouth (e.g., rose hip).

hybrid: a cross between two species.

hybridize: breeding together different species or varieties of plants or animals; the resulting hybrid often has characteristics of both parents.

inflorescence: flower cluster.

involucre: a set of bracts closely associated with one another, encircling and immediately below a flower cluster.

lanceolate: a long leaf that is widest at the middle and pointed at the tip.

lenticel: a slightly raised pore on root, trunk or branch bark.

mesic: habitat with intermediate moisture levels—not too dry or too moist.

montane: mountainous habitat, below the timberline.

multiple fruit: ripens from a number of separate flowers that grow closely together, each with its own pistil (e.g., mulberry, fig).

node: the place where a leaf or branch is attached.

nutlet: a small, hard, dry, one-seeded fruit or part of a fruit; does not split open.

opposite: situated across from each other at the same node (not alternate or whorled); or situated directly in front of another organ (e.g., stamens opposite petals).

ovary: the part of the pistil that contains the ovules.

ovules: the organs that develop into seeds after fertilization.

GLOSSARY

palmate: divided into three or more lobes or leaflets diverging from a common point, like fingers on a hand.

peduncle: a flower or fruit stem.

pemmican: a mixture of finely pounded dried meat, fat and sometimes dried fruit.

perennial: a plant that lives for three or more years, usually flowering and fruiting for several years.

pericarp: the part of a fruit that derives from the ovary wall; generally consists of three layers: (from inside to outside) endocarp, mesocarp, exocarp.

petal: a unit of the corolla; usually brightly coloured to attract insects.

phytomedicine: the use of plants as medicine.

pinnate: with branches, lobes, leaflets or veins arranged on both sides of a central stalk or vein; feather-like.

pistil: the female part of the flower, composed of the stigma, style and ovary.

pitch: sticky tree sap, for example from a pine tree.

pome: a fleshy fruit with a core (e.g., an apple) composed of an enlarged hypanthium around a compound ovary.

prostrate: growing flat along the ground.

purgative: causing watery evacuation of the bowels.

raceme: an unbranched cluster of stalked flowers on a common, elongated central stalk, blooming from the bottom up.

receptacle: an expanded stalk tip at the centre of a flower, bearing the floral organs or the small, crowded flowers of a head.

recurved: curved under (usually referring to leaf margins).

rhizome: an underground, often lengthened stem; distinguished from the root by the presence of nodes and buds or scale-like leaves.

saponin: any of a group of glycosides with steroid-like structure; found in many plants; causes diarrhea and vomiting when taken internally but commercially used in detergents.

sepal: one segment of the calyx; usually green and leaf-like.

spore: a reproductive body composed of one or several cells that is capable of asexual reproduction (doesn't require fertilization).

sporophyll: a spore-bearing leaf; a scale of a conifer cone.

spp.: abbreviation of "species" (plural).

spur: a hollow appendage on a petal or sepal, usually functioning as a nectary.

spur-shoot: a slow-growing, much-reduced shoot (e.g., on a larch or ginko tree).

stolon: a slender, prostrate, spreading branch, rooting and often developing new shoots and/or plants at its nodes or at the tip.

style: the part of the pistil connecting the stigma to the ovary; often elongated and stalk-like.

subalpine: just below the treeline, but above the foothills.

sucker: a shoot not originating from a seed, but from a rhizome or root.

tepal: a sepal or petal, when these structures are not easily distinguished.

throat: the opening into a corolla tube or calyx tube.

toxic: a substance that can cause damage, illness or death.

tundra: a habitat in which the subsoil remains frozen year-round characterized by low growth and lacking in trees.

unarmed: without prickles or thorns.

variety: a naturally occurring variant of a species; below the level of subspecies in biological classification.

Section of a regular flower with numerous carpels

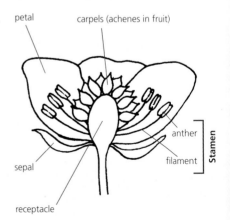

References

Bennett, Jennifer, Ed. 1991. *Berries: A Harrowsmith Gardener's Guide*. Camden House Publishing, Camden East, Ontario.

Black, Marmalade. 1977. *It's the Berries*. Hancock House Publishers Ltd, Saanichton, British Columbia.

Derig, Betty B., Margaret C. Fuller. 2001. *Wild Berries of the West*. Mountain Press Publishing Company, Missoula, Montana.

Domico, Terry. 1979. *Wild Harvest: Edible Plants of the Pacific Northwest*. Hancock House Publishers Ltd., Saanichton, British Columbia.

Elias, Thomas S., & Peter A. Dykeman. 1990. *Edible Wild Plants: A North American Field Guide*. Sterling Publishing Company Inc., New York, New York.

Hamersley Chambers, Fiona. 2011. *Wild Berries of British Columbia*. Lone Pine Publishing, Edmonton, Alberta.

Harrington, H. D. 1967. *Western Edible Wild Plants*. The University of New Mexico Press, Albuquerque, New Mexico.

Hutchens, Alma R. 1991. *Indian Herbalogy of North America: The Definitive Guide to Native Medicinal Plants and Their Uses*. Shambhala Publications Inc., Boston, Massachusetts.

Johnson, D., L. Kershaw, A. MacKinnon, & J. Pojar. 1995. *Plants of the Western Forest: Alberta, Saskatchewan & Manitoba Boreal and Aspen Parkland*. Lone Pine Publishing, Edmonton, Alberta.

Kershaw, Linda. 2003. *Alberta Wayside Wildflowers*. Lone Pine Publishing, Edmonton, Alberta.

Kershaw, Linda. 2000. *Edible & Medicinal Plants of the Rockies*. Lone Pine Publishing, Edmonton, Alberta.

Kindschner, Kelly. 1987. *Edible Wild Plants of the Prairie: An Ethnobotanical Guide*. University Press of Kansas, Lawrence, Kansas.

Kuhnlein, Harriet V., & Nancy J. Turner. 1991. *Traditional Plant Foods of Canadian Indigenous Peoples: Nutrition, Botany and Use*. Gordon and Breach Science Publishers, Philadelphia, Pennsylvania.

MacKinnon, A., L. Kershaw, J. T. Arnason, P. Owen, A. Karst, & F. Hamersley Chambers. 2009. *Edible & Medicinal Plants of Canada*. Lone Pine Publishing, Edmonton, Alberta.

MacKinnon, A., J. Pojar, & R. Coupé, Eds. 1999. *Plants of Northern British Columbia*. Lone Pine Publishing, Edmonton, Alberta.

Marles, R. J., C. Clavelle, L. Monteleone, N. Tays, & D. Burns. 2000. *Aboriginal Plant Use in Canada's Northwest Boreal Forest*. UBC Press, Vancouver, British Columbia.

Mikisew Cree First Nation. 2011. *Sagow Pimachiwin: Plants and Animals used by Mikisew Cree First Nation for Food, Medicine and Materials. Public Version*. Mikisew Cree First Nation Government Industry Relations, Alberta.

Moerman, Daniel E. 1998. *Native American Ethnobotany*. Timber Press, Portland, Oregon.

Marrone, Teresa. 2009. *Wild Berries & Fruits Field Guide: Minnesota, Wisconsin and Michigan*. Adventure Publications Inc., Cambridge, Minnesota.

Neal, Bill. 1992. *Gardener's Latin: A Lexicon*. Algonquin Books of Chapel Hill, Chapel Hill, North Carolina.

Parish, R., R. Coupé, & D. Lloyd, Eds. 1996. *Plants of Southern Interior British Columbia and the Inland Northwest*. Lone Pine Publishing, Edmonton, Alberta.

REFERENCES

Royer, F., & R. Dickinson. 2007. *Plants of Alberta: Trees, Shrubs, Wildflowers, Ferns, Aquatic Plants & Grasses.* Lone Pine Publishing, Edmonton, Alberta.

Schofield, Janice J. 1989. *Discovering Wild Plants: Alaska, Western Canada, the Northwest.* Alaska Northwest Books, Seattle, Washington.

Sept, J. Duane. 2005. *Wild Berries of the Northwest: Alaska, Western Canada, and the Northwestern United States.* Calypso Publishing, Sechelt, British Columbia.

Stark, Raymond. 1981. *Guide to Indian Herbs.* Hancock House Publishers Ltd., North Vancouver, British Columbia.

Tilford, Gregory L. 1997. *Edible and Medicinal Plants of the West.* Mountain Press Publishing Company, Missoula, Montana.

Turner, Nancy J. 2005. *The Earth's Blanket: Traditional Teachings for Sustainable Living.* Douglas & McIntyre, Vancouver, British Columbia.

Turner, Nancy J. 1997. *Food Plants of Interior First Peoples.* UBC Press, Vancouver, British Columbia.

Turner, Nancy J., & Adam F. Szczawinski. 1988. *Edible Wild Fruits and Nuts of Canada.* Fitzhenry & Whiteside, Markham, Ontario.

Underhill, J. E. (Ted). 1974. *Wild Berries of the Pacific Northwest.* Hancock House Publishers, Saanichton, British Columbia.

Vance, F. R., J. R. Jowsey, J. S. McLean and F. A. Switzer. 1999. *Wildflowers Across the Prairies.* Greystone Books, Toronto, Ontario.

Wilkinson, Kathleen. 1990. *Trees and Shrubs of Alberta.* Lone Pine Publishing, Edmonton, Alberta.

Internet Sources

Canadian Biodiversity:
http://canadianbiodiversity.mcgill.ca/english/species/plants/index.htm

Evergreen Native Plant Database:
http://nativeplants.evergreen.ca/

Flora of North America, from the Flora of North America Association:
http://www.fna.org/FNA

Natureserve:
http://natureserve.org/explorer/

United States Department of Agriculture Plants Database: Natural Resources Conservation Service:
http://plants.usda.gov/

Red swamp currant (*Ribes triste*)

Index to Common and Scientific Names

Entries in **boldface** type refer to the primary species accounts.

A

Actaea
  arguta. See *A. rubra*
  eburnea. See *A. rubra*
  rubra, 166–167
Amelanchier
  alnifolia, 72–75
  florida. See *A. alnifolia*
Anacardiaceae, 36–39, 154–155
apple, thorn. See hawthorn, black
Aralia spp., 122–123
  hispida, 123
  nudicaulis, 123
  racemosa, 123
Araliaceae, 122–123, 156–157
Arctostaphylos spp. 94–97
  alpina, 95, 119
    var. *rubra*, 96
  uva-ursi, 96
Arctuous
  alpina. See *Arctostaphylos alpina*
  rubra. See *Arctostaphylos uva-ursi*
asters, 152
athikimin. See currants

B

bake-apple. See cloudberry
baneberry
  common. See red b.
  red, 166–167
barberry, common, 124–125
bearberry, 94–97, 152
  alpine, 95
  black alpine. See alpine b.
  common, 96
  red, 96
Berberidaceae, 124–125
Berberis vulgaris, 124–125
bilberry. See also whortleberry
  bog. See blueberry, bog
  common. See whortleberry
  low. See whortleberry
bittersweet. See also nightshade
  American, 158–159
  European, 168, 169
  black bear berry. See twisted-stalks
  blackberry, dwarf red. See raspberry, trailing
blueberry, 108–113, 118
  bog, 112
  dwarf, 111
  lowbush, 111
  oval-leaved, 111
  velvet-leaved, 111
  bramble, five-leaved. See raspberry, trailing wild
  buckbrush. See snowberry, western
buckthorn, 40–41
  alder-leaved, 40, 41
  European, 40, 41
buffaloberry & soapberry, 52–53
buffaloberry
  russet. See soapberry
  silver, 53
bunchberry, 130–131
bush cranberry, 46–49
  American, 47, 48
  high, 47–48, 104, 112
  low. See high b.; lingonberry

C

Cactaceae, 126–127
cactus, prickly-pear, 126–127
  brittle prickly-pear, 127
  plains prickly-pear, 127
Caprifoliaceae, 42–49; 160–163
Celastraceae, 158–159
Celastrus scandens, 158–159
Chenopodiaceae, 126–127
Chenopodium
  capitatum, 128–129
  pallidicaule, 128
  quinoa, 128
cherry
  bird. See pin c.
  fire. See pin c.
  Pennsylvania. See pin c.
  pin, 68, 69
  red, 68–69
  sand, 69
  wild. See chokecherry
chokecherry, 70–71
clintonia & queen's cup, 132–133
Clintonia spp., 132–133
  borealis, 133
  uniflora, 133
clintonia
  one-flowered. See queen's cup
  yellow, 133
cloudberry, 92–93
Comandra livida. See *Geocaulon lividum*
comandra, northern, 152–153
Convallaria majalis, 140
Cornaceae, 50–51, 130–131
Cornus
  alba. See *C. sericea*
  canadensis, 130–131
  sericea, 50–51
  stolonifera. See *C. sericea*
  unalaschensis. See *C. canadensis*
cowberry. See lingonberry
cranberry, 104–107, 112, 152
  American bush, 47, 48
  bog, 106

INDEX

bush, 46–49
high bush, 47–48, 104, 112
low bush. See high bush c.; lingonberry
mountain. See lingonberry
rock. See lingonberry
small. See bog c.
Crataegus spp., 56–59
  chrysocarpa, 58
  columbiana, 58
  douglasii, 59
  succulenta, 59
crowberry, black, 98–99
Cupressaceae, 30–35
curlew berry. See crowberry, black
currant, 76–81
  black, 77–78
  bristly black. See prickly c.
  golden, 77, 78
  northern black, 77, 79
  northern red. See red swamp c.
  prickly, 77, 79, 82
  red swamp, 80
  skunk, 76, 78–79
  sticky, 79, 80
  trailing black, 79–80

D
devil's club, 156–157
dewberry. See raspberry, trailing
Disporum
  hookeri. See *Prosartes hookeri*
  trachycarpum. See *Prosartes trachycarpa*
dogwood
  Canada. See bunchberry
  dwarf. See bunchberry
  red-osier, 50–51
  western. See red-osier d.

E
Echinopanax horridum. See *Oplopanax horridus*
Elaeagnaceae, 52–55
Elaeagnus commutata, 54–55
elderberry, 42–45
  black, 45
  blue, 44
  red, 44–45
  Rocky mountain. See black e.
Empetraceae, 94–97
Empetrum nigrum, 98–99
Ericaceae, 98–119
Exobasidium vaccinii, 116

F
fairybells, 134–135
  Hooker's, 134, 135
  Oregon. See Hooker's f.
  rough-fruited, 135
false-hellebore, green, 139
false-wintergreen, 100–103
  alpine, 102–103
  hairy, 101
Fragaria spp., 148–151
  vesca, 151
  virginiana, 151
frog berry. See twisted-stalks; currants

G
Gaultheria spp., 100–103
  hispidula, 101
  humifusa, 102–103
  procumbens, 103
Geocaulon lividum, 152–153
gooseberry, 82–85
  Canadian. See northern g.
  northern, 85
  smooth. See northern g.
  swamp. See currant, prickly
  white-stemmed, 85

goosefoot, blite. See strawberry blite
grape
  frost. See riverbank g.
  riverbank, 120–121
Grossulariaceae, 76–85
grouseberry, 106

H
hawthorn, 56–59, 68
  black, 59
  fireberry, 58
  fleshy, 59
  red, 58
honeysuckle, 162–163
  bracted. See twinflower h.
  limber. See twining h.
  red. See twining h.
  twinflower, 163
  twining, 163
huckleberry, 112, 114–117
  black, 112, 114, 115
  black mountain. See black h.
  false, 116
  thinleaf. See black h.
  western. See blueberry, bog

J
juneberry. See saskatoon
juniper, 30–35
  common, 33–34
  ground. See common j.
  creeping, 34
  Rocky mountain, 34
Juniperus spp., 30–35
  communis, 33–34
  horizontalis, 34
  scopulorum, 34

K
kañiwa, 128
kawiskowimin, 131
kinnikinnick. See bearberry, common

180

INDEX

L
Liliaceae, 132–147
lily-of-the-valley
European, 140
feathery false. *See*
Solomon's-seal, false
starry false. *See*
Solomon's-seal, starflowered false
three-leaved false. *See*
Solomon's-seal, threeleaved
wild, 140–141
lily, yellow bluebead. *See*
clintonia, yellow
lingonberry, 105, 107
Lonicera spp., 162–163
dioica, 163
involucrata, 163

M
mahogany, mountain. *See*
yew, Pacific
Maianthemum spp.,
140–145
canadense, 140–141
racemosum, 144
stellatum, 144
trifolium, 144
matrimony vine. *See*
nightshade
mayflower, Canada. *See*
lily-of-the-valley, wild
meadow rue, western, 54
mealberry. *See* bearberry,
common
Menispermaceae, 164–165
Menispermum canadense,
164–165
Menziesia ferruginea, 116
misaskatomina. *See*
saskatoon
moonseed, Canada,
164–165
mooseberry. *See*
cranberry, high bush
moss berry. *See* crowberry,
black

mountain ash, 60–61
European, 60, 61
Greene's. *See* western m.
western, 61

N
nagoonberry. *See*
raspberry, Arctic
dwarf
dwarf. *See* raspberry,
wild red
nannyberry, 48
nightshade, 168–169
black, 169
climbing. *See* bittersweet,
European
woody. *See* bittersweet,
European

O
Oplopanax horridus,
156–157
Opuntia spp., 126–127
fragilis, 127
polyacantha, 127
owl berry. *See* twistedstalks
Oxycoccus
microcarpus. *See*
Vaccinium oxycoccos
oxycoccos. *See Vaccinium oxycoccos*
quadripetalus. *See*
Vaccinium oxycoccos

P
Parthenocissus
quinquefolia, 172, 173
vitaceae, 172–173
partridge berry. *See*
lingonberry
plum, 66–67
American, 66, 67
Canada, 66, 67
wild. *See* American p.;
Canada p.
**poison-ivy, western,
154–155**

Polygonatum biflorum,
146–147
**prickly-pear cactus,
126–127**
brittle, 127
plains, 127
Prosartes spp., 134–135
hookeri, 135
trachycarpa, 135
Prunus spp., 66–71
americana, 67
nigra, 67
pensylvanica, 69
pumila, 69
virginiana, 70–71
Psoralea esculenta, 72

Q
queen's cup, 133. *See also*
clintonia & queen's cup
quinoa, 128

R
Ranunculaceae, 166–167
raspberry, 86–89
American red. *See* wild
red r.
Arctic dwarf, 87
creeping. *See* trailing
wild r.
strawberry leaf. *See*
trailing wild r.
trailing, 88
trailing wild, 87, 88
wild red, 87
Rhamnaceae, 40–41
Rhamnus spp., 40–41
alnifolia, 41
cathartica, 41
Rhus
glabra, 36–37
radicans var. *rydbergii*.
See Toxicodendron rydbergii
trilobata, 38–39
Ribes spp., 76–85
aureum, 78
glandulosum, 78–79

181

INDEX

hirtellum. See *R. oxyacanthoides*
hudsonianum, 79
idaeus, 87
inerme, 85
lacustre, 79, 82
laxiflorum, 79–80
oxyacanthoides, 85
pedatus, 88
propinquum, 80
pubescens, 88
setosum. See *R. oxyacanthoides*
strigosis. See *R. idaeus*
triste, 80
viscosissimum, 80
Rosa spp., 62–65
acicularis, 64
alcea. See *R. woodsii*
arkansana, 64
bourgeauiana. See *R. acicularis*
suffulta. See *R. woodsii*
woodsii, 65
Rosaceae, 56–75, 86–93, 148–151
rose
Arkansas, 64
prairie, 65. See also Arkansas r.
prickly, 64
wild, 62–65. See also prickly r.
Woods'. See prairie r.
Rowan tree. See mountain ash, European
Rubus spp., 86–93
acaulis. See *R. arcticus*
arcticus, 87
chamaemorus, 92–93
parviflorus, 90–91

S

Sambucus spp., 42–45
cerulea. See *S. nigra* ssp. *cerulea*
glauca. See *S. nigra* ssp. *cerulea*
nigra ssp. *cerulea,* 44
racemosa, 44–45
var. *melanocarpa,* 45
var. *pubens,* 45
sandberry. See bearberry, common
Santalaceae, 152–153
sarsaparilla, 122–123
hairy, 123
wild, 123
saskatoon, 72–75
serviceberry. See saskatoon
Canada. See saskatoon
shadbush. See saskatoon
Shepherdia spp., 52–53
argentea, 53
canadensis, 53
silverberry, 54–55
skunkbush, 38–39
Smilacina
racemosa. See *Maianthemum racemosum*
stellata. See *Maianthemum stellatum*
trifolia. See *Maianthemum trifolium*
snake berry. See baneberry, red; fairybells
snowberry, 160–161
common, 161
creeping. See false-wintergreen, hairy.
thin-leaved. See snowberry, common
western, 161
soapberry, 53. See also buffaloberry & soapberry
Solanaceae, 168–169
Solanum spp., 168–169
americanum var. *nodiflorum,* 169
dulcamara, 169

Solomon's-seal
false, 142–145
great. See smooth S.
smooth, 146–147
star-flowered false, 142, 144
three-leaved, 144
soopolallie. See soapberry
Sorbus spp., 60–61
aucuparia, 61
scopulina, 61
spikenard, 123
spinach, strawberry. See strawberry blite
squashberry. See cranberry, high bush
strawberry, 148–151, 152
blueleaf. See wild s.
wild, 151
wood, 151
strawberry blite, 128–129
strawberry spinach. See strawberry blite
Streptopus spp., 136–139
amplexifolius, 138
lanceolatus, 138
roseus. See *S. lanceolatus*
sumac, smooth, 36–37
Svida sericea. See *Cornus sericea*
Symphoricarpos spp., 160–161
albus, 161
occidentalis, 161

T

Taxaceae, 170–171
Taxas
canadensis, 170
brevifolia, 170–171
Thalictrum occidentalis, 54
thimbleberry, 90–91
toadflax, false. See comandra, northern
Toxicodendron rydbergii, 154–155
turnip, prairie, 72

GLOSSARY

twinberry, black. *See* honeysuckle, twinflower
twisted-stalk, 136–139
  clasping, 138
  rosy, 138

U
Uva corinthiaca, 77

V
Vaccinium spp., 104–119
  angustifolium, 111
  caespitosum, 111
  membranaceum, 115
  myrtilloides, 111
  myrtillus, 118–119
  occidentale. See V. uliginosum
  oreophilum. See V. myrtillus
  ovalifolium, 111

oxycoccos, 106
scoparium, 106
uliginosum, 112
vitis-idaea, 107
Viburnum spp., 46–49
  edule, 47–48, 112
  lentago, 48
  opulus, 48
  trilobum var. *americanum. See V. opulus*
Virginia creeper, 172, 173
  false, 172–173
Vitaceae, 120–121, 172–173
Vitis riparia, 120–121

W
whortleberry, 115, **118–119.** *See also* bearberry, alpine

willow
  red. *See* dogwood, red-osier
  wolf. *See* silverberry
wintergreen, 103
  alpine. *See* false-wintergreen, alpine
  creeping. *See* false-wintergreen, alpine
witch berry. *See* twisted-stalks
woodbine. *See* Virginia creeper, false

Y
yew
  Canada, 170
  Pacific, 170–171
  western. *See* Pacific y.

Z
zereshk, 124

Photo & Illustration Credits

Photo and illustration numbers refer to page number and location letter on the page. Page location is identified by sequential letters (a, b, c, d, etc.) for each page, running top to bottom on the left column, then top to bottom on the right column. Photo and illustration location letters are identified separately.

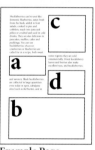

Example Page

Photos
Lee Beavington: 15, 17, 18a, 22, 25, 26, 27b, 28a, 32, 33a, 33c, 42, 43a, 43b, 44, 45a, 45b, 46, 47, 49d, 50, 51, 56, 62, 70, 72, 74, 75a, 75b, 76, 87a, 90, 95, 97c, 107a, 107b, 108, 112, 113, 114, 130, 131, 132, 135, 136, 137a, 138a, 139, 142, 143, 145a, 149, 151, 156, 157, 160, 166, 167a, 167b, 168, 170. **Todd Boland:** 40, 59, 110, 124, 133, 140, 141. **Paul S. Drobot:** 146, 147. **Tracy Ruta Fuchs:** 66, 67, 126. **Fiona Hamersley Chambers:** 184a. **Cory Harris:** 18b, 21, 41, 49a, 120, 121, 165, 172. **Neil Jennings:** 54, 55a. **Amanda Karst:** 184b. **Linda Kershaw:** 20, 27a, 33b, 34, 37, 38, 39, 52, 60, 68, 69b, 80, 81a, 81b, 87b, 94, 97a, 98, 102, 122, 129, 134, 137b, 138b, 145b, 150a, 150b, 153b, 153c, 153d, 155b, 178. **Ron Long:** 55b. **Robin Marles:** 145c, 158, 161, 164. **Virginia Skilton:** 24, 28b, 36, 82, 84, 103, 154, 162. **Superior National Forest:** 153a. **Thinkstock:** 1. **Robert D. Turner and Nancy J. Turner:** 12, 13, 19, 29, 30, 31, 35, 49b, 49c, 57, 63, 65, 78, 85, 86, 88, 92, 93, 97b, 99, 100, 104, 111, 116, 117, 118, 148, 155a, 169, 173. **Per Verdonk:** 58a, 69a, 128.

Illustrations
Frank Burman: 31a, 31b, 34, 37, 39b, 41, 44a, 47, 48b, 48c, 48d, 53a, 57, 58a, 59a, 59b, 61a, 64a, 64c, 67a, 67b, 69a, 69b, 77, 78, 79a, 80a, 80b, 81a, 85, 86, 88a, 88b, 89a, 89b, 93a, 93b, 96a, 96b, 96c, 99, 101, 102b, 105, 106b, 109a, 109b, 110, 111, 112, 119, 121, 122, 123a, 123b, 125, 127a, 127b, 129, 133d, 135a, 135d, 139a, 141a, 141b, 143a, 143b, 144b, 149, 150, 155, 157, 159a, 159b, 161, 163b, 169a, 169b. **Linda Kershaw:** 14, 15a, 15b, 16a, 16b, 16c, 174, 176. **George Penetrante:** 40, 102a, 133a, 133b, 147, 152, 165, 173. **Ian Sheldon:** 35, 39a, 43, 44b, 48a, 50, 51, 53b, 55a, 55b, 58b, 61b, 61c, 64b, 71a, 71b, 71c, 73a, 73b, 79b, 81b, 83, 91a, 91b, 96d, 106a, 106c, 115a, 115b, 116, 131, 133c, 135b, 135c, 138, 139b, 144a, 151, 160, 163a, 167, 171

About the Authors

FIONA HAMERSLEY CHAMBERS was born in Vancouver. She holds an undergraduate degree from the University of Victoria in French and Environmental Studies (1994), a Masters of Science in Environmental Change and Management from Oxford University (1998), and a Masters in Environmental Design from the University of Calgary (1999). Speaking English, French and Spanish, she has travelled extensively throughout Europe, Australia, New Zealand, Mexico and Central America and has a strong interest in learning about traditional plant uses wherever she goes. Fiona has taught Environmental Studies at the University of Victoria since 1999, and ethnobotany at Pacific Rim College since 2009. She currently divides her time between teaching, running a small organic farm and food plant nursery (www.metchosinfarm.ca), writing books and academic papers, and raising two energetic boys who also love plants, animals and bugs.

AMANDA KARST was born in Regina and has spent most of her life on the prairies. She obtained a Bachelor of Science at the University of Regina and a Masters of Science in Biology, with a focus on plant ecology and ethnobotany, from the University of Victoria. She has worked on a range of projects in ethnobotany and plant ecology in Alberta, Saskatchewan, Quebec and Newfoundland and Labrador. She now resides in Winnipeg with her husband and daughter and loves local food, canoe trips and berry picking. She has worked as a Research Associate at the Centre for Indigenous Environmental Resources since 2006.